BLENDED CEMENTS

A symposium
sponsored by
ASTM Committee C-1
on Cement
Denver, CO, 27 June 1984

ASTM SPECIAL TECHNICAL PUBLICATION 897
Geoffrey Frohnsdorff, National Bureau of
Standards, editor

ASTM Publication Code Number (PCN)
04-897000-07

 1916 Race Street, Philadelphia, PA 19103

Library of Congress Cataloging-in-Publication Data

Blended cements.

(ASTM special technical publication; 897)
"ASTM Symposium on Blended Cements sponsored by Committees C-1 on Cement—Introd.
"ASTM publication code number (PCN) 04-897000-07."
Includes bibliographies and index.
1. Cement—Congresses. 2. Aggregrates (Building materials)—Congresses.
I. Frohnsdorff, Geoffrey. II. ASTM Symposium on Blended Cements (1984: Denver, Colo.) III. ASTM Symposium on Blended Cements (1984: Denver, Colo.)
IV. American Society for Testing and Materials. Committee C-1 on Cement. V. Series.
TA434.B57 1986   666'.94   85-28586   ISBN 0-8031-0453-7

NOTE

The Society is not responsible, as a body,
for the statements and opinions
advanced in this publication.

Printed in Ann Arbor, MI
January 1986

Foreword

The symposium on Blended Cements was held in Denver, Colorado, on 27 June 1984. ASTM Committee C-1 on Cement sponsored the symposium. Geoffrey Frohnsdorff, National Bureau of Standards, served as symposium chairman and editor of this publication.

Related
ASTM Publications

Masonry: Materials, Properties, and Performance, STP 778
(1982), 04-778000-07

Cement Standards–Evolution and Trends, STP 663 (1979),
04-663000-07

A Note of Appreciation to Reviewers

The quality of the papers that appear in this publication reflects not only the obvious efforts of the authors but also the unheralded, though essential, work of the reviewers. On behalf of ASTM we acknowledge with appreciation their dedication to high professional standards and their sacrifice of time and effort.

ASTM Committee on Publications

ASTM Editorial Staff

Helen M. Hoersch
Janet R. Schroeder
Kathleen A. Greene
Bill Benzing

Contents

Introduction

Blended cements are usually, but not always, blends of portland cements with other finely ground materials. The most common ingredients for blending with portland cements are pozzolans and latent hydraulic materials, such as, ground granulated blast-furnace slags, but other materials, such as ground limestone, may be also used. Blended cements without portland cement are sometimes made, an example being slag cements made from ground, granulated blast-furnace slag and slaked lime or gypsum.

Blended cements have been manufactured in many countries but, at least in the United States, the volumes manufactured have been small compared to the volumes of portland cements. A renewed interest in blended cements came about in the United States following the oil embargo in 1973. This was because portland cement manufacture is energy-intensive and blended cements generally require less energy per unit volume to manufacture. The ASTM specifications for blended cements which existed in 1973 were not as well-developed as the portland cement specifications. This probably was the result of less interest and greater complexity in defining the product. The blended cement standards appeared to be too restrictive in terms of the ingredients permitted and in the range of acceptable proportions. However, in the absence of adequate data on factors affecting the performance of blended cements, the specifications have been difficult to change.

This volume presents the papers which were presented at the ASTM Symposium on Blended Cements sponsored by Committees C-1 on Cement. The Symposium was organized to provide more information on blended cements of all kinds, so as to aid the standards development process. We hope and believe it will achieve its purpose.

Geoffrey Frohnsdorff
National Bureau of Standards, Washington, DC
20234; symposium chairman and editor.

1

Portland Blast-Furnace Slag Cements

Jean Daube [1] and Robert Bakker [2]

Portland Blast-Furnace Slag Cement: A Review

REFERENCE: Daube, J. and Bakker, R., **"Portland Blast-Furnace Slag Cement: A Review,"** *Blended Cement, ASTM STP 897,* G. Frohnsdorff, Ed., American Society for Testing and Materials, Philadelphia, 1986, pp. 5–14.

ABSTRACT: It is a two-part report; in the first section the properties required for the components of blast-furnace slag cements, cement standards, and applications are described; in the second section, a review of the performance of blast-furnace slag cements compared to portland cement is described with special reference to sulfate and seawater resistance, reduced alkali-silica expansion, and low-heat properties.

KEY WORDS: specifications, use, portland slag cements, properties, slags, performance, cements, standards, sulfate resisting cements, seawater corrosion, alkali silica expansion, low-heat cements

Blast-furnace slag cements (BFSC) have been used for decades in Europe and in many cases they are used to replace normal portland cements (OPC). In the BENELUX market (Belgium, Netherlands, and Luxembourg), which annually amounts to a total of 9 million metric ton of cements, approximatively 50% are BFC.

Quality of BFC Components

Quality of Iron Blast-Furnace Slag

Two main characteristics determine the hydraulic properties of granulated iron blast-furnace slag (BFS), that is, its chemical composition and its vitreous state.

Based on the results of a previous study [1], an empirical formula defines an hydraulic index I_h which characterizes the quality of the slag

$$I_h = \frac{CaO + 1.4\,MgO + 0.56\,Al_2O_3}{SiO_2}$$

[1] Research manager, S. A. Cimenteries CBR, Brussels, Belgium.
[2] Fellow-worker, Vereniging Nederlandse Industrie, Hertogenbosch, The Netherlands.

TABLE 1 — *Hydraulic values of BFS from chemical composition.*

Hydraulic Index I_h	Expected Quality of Slag
<1.65	unacceptable
1.65 to 1.85	normal
1.85 to 2.10	superior

This formula was obtained by correlating the hydraulic index of 56 laboratory slags with the mechanical strengths of cement mortars made from those slags and which were tested after 3 and 7 days in accordance with the European ISO method on 4 by 4 by 16-cm mortar bars. The validity of this formula was confirmed recently by tests on industrial slags [2].

The fact that the coefficients affecting magnesia and alumina are identical, respectively, to the mole ratios calcium oxide/magnesium oxide (CaO/MgO) and calcium oxide/aluminum oxide (CaO/Al$_2$O$_3$), denotes a molecular substitution of CaO in the vitreous slag; silicon-ion is to be considered as a network former, and calcium, magnesium, and aluminum ions as network modifiers [2].

Statistical data for industrial slags from BENELUX reveal I_h-values ranging from 1.55 to 2.10. Table 1 gives a rough classification of the hydraulic quality of slags within this range.

In order to have hydraulic properties, the slag must have been quenched and fixed in its vitreous state which may be checked by means of X-ray diffraction analysis. Systematic X-ray analysis is generally not required but may be important for the products obtained from new quenching methods such as, for example, expanded pelletizing. A complete vitrification is not required, and a low percentage of crystallization (3 to 5%) may improve the reactivity [2].

It is known that magnesia does not appear in granulated slags in its crystallized form, periclase, but enters as a modifier in the silicon network of the glass. Consequently, high magnesia contents in the slags will not induce expansion in concrete.

A direct measurement of the compressive strength on industrial cements replaces the slag activity test described in the ASTM Specification for Ground Iron Blast-Furnace Slag for Use in Concrete and Mortars (C 989-82).

Quality of Portland Cement Clinker and Calcium Sulfate

Intergrinding of portland clinker, BFS, and calcium sulfate (CaSO$_4$) allows the three components to adapt to each other. The reactivity of slag is enhanced by alkalinity; experience has confirmed that clinker with a lime saturation factor (LSF) in excess of 0.95 and a relatively high alkali-oxides and -sulfates content will favor the early strength development of the BFSC. Consequently, clinker from dry-process kilns is a better activator for BFS because it tends to have higher alkali content.

TABLE 2—*Maximum SO_3 limits in cement standards in the United States, United Kingdom, and BENELUX countries.*

Slag in Cement, %	Maximum SO_3 Allowed In				
	USA	UK	Belgium	Netherlands	Luxembourg
27 to 70	3.0	3.0	3.75	4.0	4.5
$\geq$ 70	4.0	3.0	3.75	4.0	4.5
$\geq$ 85	4.0	3.0	5.00	$\cdots$	4.5

The nature of $CaSO_4$ plays an important role in the regulation of the setting of BFSC. Special attention must be paid when using a $CaSO_4$ waste product such as phospho-gypsum or anhydrite. This waste sulfate may contain traces of retarding constituents, the effects of which will be proportional to the slag percentage in the cement. The choice between gypsum and anhydrite as a regulator is of utmost importance so as to avoid rheological difficulties during handling of concrete. It has been proven that BFSC with anhydrite will sometimes result in quick set, especially in the presence of admixtures, while slag cement interground with gypsum only may result in false set. So it is often advisable to produce BFSC using a mixture of both gypsum and anhydrite. The ratio $CaSO_4 \cdot 2H_2O/CaSO_4$ will be adjusted by the cement manufacturer as a function of the clinker content and the clinker composition.

An important and disputable problem stems from the maximum admissible sulfur trioxide (SO_3) content in BFSC. $CaSO_4$ favorably influences the strength development of BFSC, and it would be detrimental to limit unnecessarily its content because of exaggerated fear of the product's not meeting the requirements of volume stability.

Table 2 gives a comparison between the SO_3 requirements in ASTM, British, and the corresponding standards in the BENELUX countries.

Due to the accelerating properties of SO_3, an increase in its limits as set forth in the United States and British Standards would allow to improve the quality of BFSC, especially in the case of cements with a high percentage of slag.

Performance of Blast-Furnace Slag Cement

Table 3 compares the ASTM, British, and BENELUX denominations of BFSC. In addition to this nomenclature, in the BENELUX distinction is made between three classes of compressive strengths, as shown in Table 4. These classes are applicable for OPC as well as for BFSC. Nevertheless because of technical or economical reasons not all types of cement are produced in every strength class. For example, there is no class 33.3 MPa (4830 psi) BFSC on the market.

The strength requirements for ASTM Types I and III on one hand, and for Types 40 (B) and 50 (C) on the other hand are rather similar. On the contrary, the

TABLE 3 — *Nomenclature of blended hydraulic cement following slag content.*

Clinker, %	ASTM	British	Belgium	Netherlands	Luxembourg	Slag, %
100						0
90	Slag-modified portland cement, Type I (SM)					10
80		portland blast-furnace cement, Type PBLF	slag portland cement, Type PL	slag portland cement, Type psc		20
70	Portland blast-furnace slag cement, Type I S		Iron portland cement, Type PF		Iron portland cement, Type PF	30
60						40
50						50
40		low heat portland blast-furnace cement, Type LH PBLF	blast-furnace cement, 35/60 Type HK, 60/85 Type HL	blast-furnace cement, Type hc	blast-furnace cement, Type HF	60
30						70
20	Slag cement, Type S					80
10			Permetallurgic cement, Type LK		Permetallurgic cement, Type PM	90
0						100

TABLE 4—Compressive strength requirements psi (MPa) of cements from BENELUX. Test method ASTM C 109.

Belgium and Luxembourg						Netherlands				
Class	Compressive Strength At					Class	Compressive Strength At			
									28 d	
N/mm²	psi,ᵃ (MPa)	1 d	3 d	7 d	28 d		1 d	3 d	min	max
30	2900 (20.0)	⋯	⋯	1550 (10.7)	2900 (20.0)	A	⋯	1060 (7.3)	3380 (23.3)	5320 (36.7)
40	3870 (26.7)	⋯	1550 (10.7)	2900 (20.0)	3870 (26.7)	B	⋯	1550 (10.7)	4350 (30.0)	6290 (43.3)
50	4830 (33.3)	1550 (10.7)	2900 (20.0)	⋯	4830 (33.3)	C	1550 (10.7)	2900 (20.0)	4830 (33.3)	⋯

ᵃEmpirical coefficients are used to convert ISO to ASTM C 109 test values:

$$\text{ASTM (psi)} = 96.7 \text{ ISO (N/mm}^2) \text{ or ASTM (MPa)} = \frac{\text{ISO (N/mm}^2)}{1.5}.$$

ASTM Type IV requirements — 6.9 and 17.2 MPa, respectively, at 7 and 28 days — are quite below those of the Belgian Type 30 low-heat cement which must reach 10.7 and 20.0 MPa, respectively, at 7 and 28 days. Experience has proven that as far as strength is concerned, ASTM Type S cement certainly has the potential to go far beyond the ASTM Specification for Portland Cement (C 150-83a) requirements.

The classification of cements by categories of strength reveals the desirability to provide for a large interchangeability between OPC and BFSC. This appears clearly when considering the results obtained in the plant laboratories. Table 5 gives an example of two pairs of cement belonging, respectively, to classes 26.7 and 20 MPa. It can be noted that the cements of each class do not only comply with the requirements as shown in Table 4, but also have a very similar strength development with time.

The contractor must follow the concrete standards, but in BENELUX countries no specification for a type of cement is set forth depending on the kind of application. Concrete standards only state that cement must comply with cement standards. The contractor is free to use either OPC or BFSC. Limitation of the choice of cement through the concrete standards may be uneconomical in as much as this is not imposed by special technical requirements. This complete inter-changeability between OPC and BFSC applies specifically in the mild and rather humid sea-climate in the BENELUX.

In BENELUX countries, BFSC have been used for decades in every field of construction, for example, multi-storied buildings, continuously reinforced concrete pavements, and dams.

From Table 5, it appears that BFSC show a tendency to develop early strength more slowly than OPC, but that later on they reach a higher level of strength. This is particularly true at temperatures below 10°C (50°F) and, in practice, influences the choice of the type of cement. On the other hand, BFSC are especially suited for steam curing.

Performance of Slag Cement

It is well known from practice that BFSC with a slag content equal or above 65% resist sulfate and seawater attack. It was demonstrated in a previous report [3] that this property is due to the very low permeability of concrete made with a slag cement. The test results given in Table 6 show that the rate of diffusion of ions in concrete made of cement with high slag contents is very slow in comparison with OPC concrete. This reduced rate of permeability can be fully explained by the fact that BFSC behave like a two-component system wherein interlocking additional hydrates precipitate into the free space between clinker and slag particles, thus filling the pores between both [4].

Therefore in the BENELUX standards sulfate resisting cements are considered to be either portland cement with a maximum of 5% Al_2O_3, 5% MgO, and 3% $3CaO \cdot Al_2O_3$, or BFSC with a slag content of 65 to 80% in the Netherlands and of 70 to 85% in Belgium.

TABLE 5—*Comparison between compressive strength values psi (MPa) of industrial cements of the same classes.*

| Class | Equivalent Symbols | | Compressive Strength At | | | | | | | | Specific Surface Blaine, m²/kg | |
	USA	Belgium	2 days avg	s	3 days avg	s	7 days avg	s	28 days avg	s	avg	s
3870 (26.7)	I	P40	1840 (12.7)	230 (1.6)	2510 (17.3)	220 (1.5)	3770 (26.0)	310 (2.1)	5030 (34.7)	280 (1.9)	342	26
	IS	HK 40	1740 (12.0)	200 (1.4)	2320 (16.0)	280 (1.9)	3670 (25.3)	300 (2.1)	5420 (37.3)	240 (1.7)	448	27
2900 (20.0)	P	PP$_z$ 30	1350 (9.3)	190 (1.3)	1740 (12.0)	250 (1.7)	2710 (18.7)	290 (2.0)	3960 (27.3)	340 (2.3)	370	32
	S	LK 30	...	...	1550 (10.7)	160 (1.1)	2510 (17.3)	150 (1.1)	3580 (24.7)	135 (0.9)	450	80

NOTES—The coefficients of conversion ISO to ASTM are the same as for Table 4.
Proportion by weight = type P 40: 100 clinker/0 slag
HK 40: 58 clinker/42 slag
PP$_z$ 30: 80 clinker/20 fly ash
LK 30: 15 clinker/85 slag.

TABLE 6—*Diffusion coefficients of sodium, potassium, and Cl-ions in hardened cement paste and mortars made with PC and BFC after different hardening times.*

Diffused Ion	Hardening Time, days	W/C Ratio	D_m in 10^{-8} cm^2/s OPC	D_m in 10^{-8} cm^2/s BFSC	Slag Content of BFC, %
Na$^+$	3	0.50	7.02	1.44	75
	14	0.50	2.38	0.10	75
	28	0.55	1.47	0.05	60
	28	0.60	3.18	0.05	60
	28	0.65	4.73	0.06	60
K$^+$	3	0.50	11.38	2.10	75
	14	0.50	3.58	0.21	75
Cl$^-$	28	0.55	3.57	0.12	60
	28	0.60	6.21	0.23	60
	28	0.65	8.53	0.41	60
	5	0.50	5.08	0.42	75
	103	0.50	2.96	0.04	75
	60	0.50	4.47	0.41	65

The low permeability explains also the good performance of BFSC against alkali-silica expansion. In spite of a voluntary addition of alkalies to BFSC no critical expansion was observed during tests conforming to ASTM Test Method for Effectiveness of Mineral Admixtures in Preventing Excessive-Expansion of Concrete Due to the Alkali-Aggregate Reaction (C 441-81) (Fig. 1). Extensive experiments in Germany have confirmed this good performance [5]. However, in our countries, no case of alkali-silica expansion was ever recorded, therefore, rendering any prescription about low alkali cement unnecessary.

A low permeability also means a low probability for chlorides to reach the concrete reinforcement.

Finally, slag cement Type S with a slag content near 85% finds an application with low hydration heat. Cements with a hydration heat of 155 kJ/kg at 3 days and 8.7 (1260), 16.7 (2420), 25.3 MPa (3680 psi), respectively, at 3, 7 and 28 days are currently made and used for mass concrete work, such as dams and harbor construction. It can be noted that such Type S cements meet the requirements of Class 20.0 MPa (2900 psi) given in Table 4, but hydrates with a slower development of heat compared with OPC of the same class.

The ASTM Standard Test Method C 186-82 or British Standard 1370: part 2: 1974 for determining the hydration heat of hydraulic cement by dissolution in a calorimeter is not applicable to BFSC, first because slag partly oxidizes during hydration and particularly during the measurement of the loss on ignition, often giving an ignition gain; in addition, the specimens of low heat cements in their early stages ages (up to 7 days) are losing moisture during handling prior to testing. The Belgian B 12-213 as well as Dutch NEN-3550 standards describe a method for determining the hydration heat by means of an isothermal conduction calorimeter similar to the one earlier developed by the Portland Cement Association [6]; the method is applicable to both OPC and BFSC. In Belgium, there are

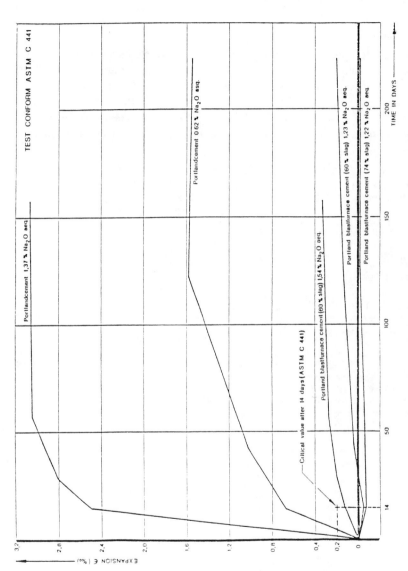

FIG. 1—*Expansion tests on OPC and BFSC conforming to ASTM C 441.*

no official requirements, but it is often specified that a low heat cement must remain below 170 kJ/kg at 3 days, 230 kJ/kg at 7 days, and 270 kJ/kg at 28 days and 20°C (68°F). In the Netherlands, a low heat cement is specified as a cement giving less than 270 kJ/kg at 7 days.

Conclusion

In order to optimize the strength and strength development of blast-furnace cement, the components slag, clinker, and $CaSO_4$ must be adjusted to each other.

With the exception of cases where high early strength is required or during winter periods with temperatures below 10°C (50°F), portland and blast-furnace cements of the same class of strength are quite interchangeable, and, in BENE-LUX countries, the choice between them is practically determined by economical considerations.

Blast furnace cements with a slag content exceeding 65% are sulfate and seawater resistant due to their very low diffusion coefficient for water and ions; for the same reason these cements are not affected by alkali-silica expansion.

Finally, slag cement with about 85% slag has been proven to be low heat with a sufficient level of strength to allow the building of dams.

References

[1] Cheron, M. and Lardinois, C. in *Proceedings*, 5th International Symposium on Chemistry of Cement, Tokyo 1968, Vol. IV, pp. 277–285.

[2] Demoulian, E., Gourdin, P., Hawthorn, F., and Vernet, C. in *Proceedings*, 7th International Congress on Chemistry of Cement, Paris 1980, Vol. II, pp. III, 89–94.

[3] Bakker, R. F. M., thesis presented on 11 June 1980 at Rheinisch-Westfälische Technische Hochschule, Aachen, West Germany.

[4] Bakker, R. F. M. in *Proceedings*, 1th International Conference on the Use of Fly Ash, Silica Fume, Slag, and Other Mineral By-Products in Concrete, Montebello 1983, Canada ACI, Publication SP-79, p. 539.

[5] *Schriftenreihe der Zementindustrie*, Vol. 40, 1973, pp. 91–97.

[6] Monfore, G. E. and Ost, B., *Journal of the Portland Cement Association*, R&D Laboratory, Vol. 8, No. 2, 1966, pp. 13–20.

Giuseppe Frigione[1]

Manufacture and Characteristics of Portland Blast-Furnace Slag Cements

REFERENCE: Frigione, G., **"Manufacture and Characteristics of Portland Blast-Furnace Slag Cements,"** *Blended Cements, ASTM STP 897,* G. Frohnsdorff, Ed., American Society for Testing and Materials, Philadelphia, 1986, pp. 15–28.

ABSTRACT: The paper gives an account of the results collected through a wide laboratory and industrial investigation carried out in order to show in which way the compressive strength and the heat of hydration are controlled by a large number of manufacture parameters, such as glass content of the slag, grinding fineness, and gypsum amount of the blended cement. Besides, the results show that the resistance to sulfates and to alkali-aggregate reaction remain very strong, if the slag percentages in the cement is high, irrespective of gypsum content and grinding fineness of the blended cement as well as of the glass amount of the slag.

KEY WORDS: portland blast-furnace slag cement, portland cement, compressive strength, sulfate resistance, alkali-aggregate reaction, heat of hydration, gypsum content, glass content, fineness

The properties of portland blast-furnace slag cements (PBFSCs) will become more and more determined by the quantity of the blast-furnace slag (BFS) used in the manufacture of these cements [*1*,*2*]. In the present study we refer to cements in which the slag is the dominant component, typically used in amounts between 60 and 70%. Such cements have a long history of successful use in all types of applications in Europe [*3*,*4*].

For construction in sulfate environments, these PBFSCs are used as an alternative to low tricalcium aluminate portland cement (PC) [*5*,*6*]. With reference to expansion of concrete made with alkali-reactive aggregates, these PBFSCs were found to represent a valid defense [*7*–*9*]. More specifically, as a precautionary measure, it was indicated that if the BFS content in the cement is at least 50%, the alkali content in the latter must be lower than 0.9% equivalent sodium oxide (Na_2O), while if the slag content exceeded or is equal to 65%, the equivalent Na_2O content in the cement may be up to 2.0% [*10*,*11*].

[1]Director of Research and Control, Cementir-Cementerie del Tirreno, Naples, Italy.

PBFSCs with 60% or more of slag are most suitable when low heat of hydration is needed [12]. In fact they allow high-strength concretes to be made, in mass construction, even with a reduced heat of hydration rate, and this evidently with notable benefit to the durability of the construction [13]. With reference to mechanical performance, it is undeniable that strength development is generally slower than in PCs at early ages; whereas, it has been determined that strength gain at later ages allows the achievement of values often higher than those for PCs [11,14,15].

The performances of these cements are controlled by a large number of significant manufacture parameters, some of these, such as: glass content of the slag [16–19], and chemical composition of the slag [11,20,21], are beyond the control of the cement producer, while others: grinding fineness [22–24], gypsum dosage [25–27], and composition of the portland clinker [21,28,29], are variable factors with which the cement producer is able both to counterbalance, as much as possible, the factors outside his control and to confer particular characteristics on the cement.

The present paper reports some results of a wide laboratory and industrial investigation carried out to determine the influence of these parameters on the performances of these cements.

Procedure

Materials

Slag from different Italian steel plants were selected. All were quenched in seawater. Normal portland cement clinkers (ASTM Type I) were selected also. Ranges of chemical data of the slags and portland clinkers are given in Table 1.

Cements all having the composition ratio 65% slag–35% portland clinker were prepared with these materials. Gypsum content is indicated each time. The gypsum had a $CaSO_4 \cdot 2H_2O$ content about 90%. When not otherwise indicated, slag and clinker were ground together. It should be noted that the slags used with different vitrosity are from high glass content slags wherein there has been partial crystallization after long stored in outdoor-heap [30].

Tests

Compressive strengths as well as some other physical and chemical properties of the cements tested were determined in accordance with ASTM Test for Compressive Strength of Hydraulic Cement Mortar (C 109-80), ASTM Test for Fineness of Portland Cement by Air Permeability Apparatus (C 204-81), and ASTM Test for Heat of Hydration of Hydraulic Cement (C 186-82). The chemical composition was determined in accordance with ASTM Test for Chemical Analysis of Hydraulic Cement (C 114-82a). Resistance to alkali-aggregate reaction was tested in accordance with ASTM Test for Effectiveness of Mineral Admixtures in Preventing Excessive Expansion of Concrete due to Alkali-

TABLE 1 — *Ranges of chemical data for slags and portland clinkers.*

SLAG	
Loss on ignition	0.2 to 1.5
SiO_2	33.6 to 37.8
Al_2O_3	8.8 to 13.1
Fe_2O_3	0.5 to 2.0
CaO	40.8 to 46.7
MgO	3.0 to 7.1
Na_2O	0.1 to 0.7
K_2O	0.1 to 0.6
S	0.6 to 1.3

PORTLAND CLINKER	
C_3S	41 to 60
C_2S	20 to 41
C_3A	2 to 10
C_4AF	8 to 13

Aggregate Reaction (C 441-81), expansion being measured in accordance with ASTM Test for Portland Alkali Reactivity of Cement-Aggregate Combinations (Mortar-Bar Method) (C 227-81). This investigation was carried out over a period of six months. The sulfate resistance of the cements were evaluated by two test procedures:

Proposed ASTM Method — This test method, proposed in 1977 by ASTM Subcommittee 001.29, measures the expansion of mortar bars stored in a solution of sodium sulfate ($NaSO_4$) and magnesium sulfate ($MgSO_4$) in which each salt is present in the amount of 0.176 mole/L of the solution. Mortar bars are prepared using mortar as described in ASTM Test Method C 109-80 and cured to 21 ± 1.0 MPa in cubes made of the same mortar, before the bars are immersed. In this manner tests begin at equivalent strength and presumably equivalent impermeability.

Mehta test — In this test the sulfate aggressive solution is periodically neutralized by dilute sulfuric acid (H_2SO_4). Microcubes (12.75 mm) of neat cement paste were prepared with a water/cement (W/C) ratio 0.50; after 24-h curing in mold at 20°C and 95% relative humidity, they were cured in water at 50°C for 6 days so as to produce suitable hydration and strength development before their immersion in 4.0% $NaSO_4$ solution (pH = 6.2) at 20 ± 1°C. To maintain the sulfate solution within narrow limits of pH, the solution was adjusted daily with 0.1 N H_2SO_4. This enabled the pH to be kept within 6.2 ± 0.5. After 28 days storage in sulfate solution specimens, six for each cement were submitted to compression test; specimens water cured at 50°C for six days were used as Ref *31*.

The glass content of BFSs (that is, degree of vitrification) was obtained by difference measuring the content of each crystalline component by quantitative X-ray diffraction. This procedure was accomplished by measuring the specific

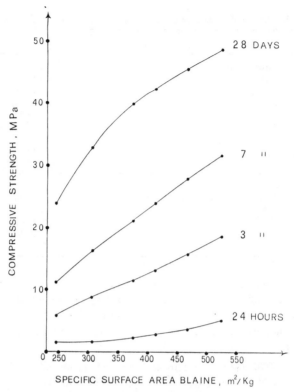

SPECIFIC SURFACE AREA BLAINE, m^2/Kg

FIG. 1 — *Compressive strength versus Blaine fineness of PBFSC.*

peak intensities of each crystalline mineral encountered relative to that of an added internal standard (10% by weight calcium fluorite (CaF_2)). The weight fraction of each mineral was then obtained by comparison of these intensity ratios in each slag to the intensity ratios for pure, synthetically produced slag minerals (C_2MS_2, C_2AS, CMS_2, C_3MS_2, CMS) [32].

Results and Discussion

Compressive Strengths

The effect of grinding fineness of PBFSC on the development of the compressive strengths is indicated in Fig. 1. For each Blaine value indicated is reported the value of compressive strength corresponding to the cement with the sulfur trioxide SO_3 content required to obtain the best strength at 3 days. For comparison, an analogous diagram is given in Fig. 2 referring to PC alone made with the same clinker. One may note the activating effect of grinding fineness on the compressive strengths of PBFSC compared to PC.

Figure 3 reports, for the same cements, in function of the Blaine fineness, the SO_3 content which the various cements required to present the best strengths at

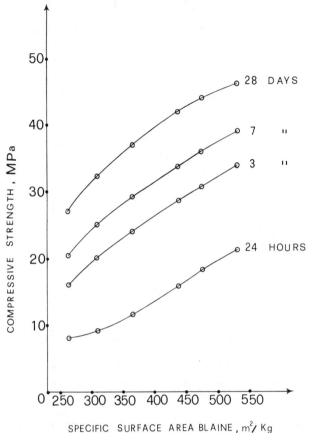

SPECIFIC   SURFACE   AREA BLAINE , m²/ Kg

FIG. 2 — *Compressive strength versus Blaine fineness of PC.*

initial 3 days and medium 28 days ages. It is noted that with the progress of age, both for PBFSC and for PC, the SO_3 content needed to reach maximum strength development decreases. However, for PC the change in SO_3 is very limited, about 0.1 percentage point, so that it can practically be considered constant, for PBFSC this variation is much more sensitive, about 0.5 percentage points, so that in practice it is necessary to decide if one prefers to have the best strengths at short or medium ages.

The effect of SO_3 content, with the varying of the degree of slag vitrosity, on the development of the compressive strengths is shown in Figs. 4, 5, and 6. The tests were made on portland clinker and slag ground separately both with various additions of gypsum and then mixed to a 35/65 clinker/slag ratio. Figures 4, 5, and 6, respectively, show the value furnished by the cements with: (*a*) slag having glass content 97%, ground to 382 m²/kg Blaine, (*b*) slag having glass content 51%, ground to 382 m²/kg Blaine, and (*c*) slag having glass content 51%, ground with a power consumption equal to that of the slags with glass content 97%.

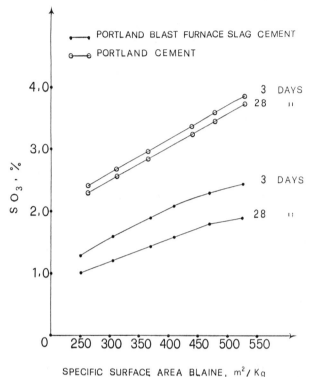

SPECIFIC SURFACE AREA BLAINE, m²/ Kg

FIG. 3 – *SO₃ content corresponding to optimum as a function of Blaine fineness.*

The fineness of portland clinker was 380 m²/kg Blaine. The lower part of the diagrams give the percentage of power consumption in the grinding, imposing 100% power consumption to grind slag with a glass content 97 and 0.90% SO_3 content. One notes, in general, the lesser sensitivity of slag with lower glass content to gypsum and also the impossibility of controlling the grinding through the Blaine fineness numbers when slags with different glass contents are used. Grinding cements at equal power consumption, one notes that with an increase in the early strength of cements produced with slags with lower glass content, there is a decrease in strengths at middle ages. One notes also that grinding at like Blaine fineness value, the power consumption is lower when the SO_3 content increases, owing to the enrichment of the gypsum in the fine fraction [33, 34].

Figure 7 shows the effect of the glass content of the slag on the development of the compressive strength of portland clinker — slag mixtures ground separately to a Blaine fineness value of 360 m²/kg. It is observed, in accordance with Demoulian and co-workers [35], that relatively small percentages of crystalline fraction in the slag can have a considerable effect on the development of the strengths. This, probably, because these small quantities of crystals are distrib-

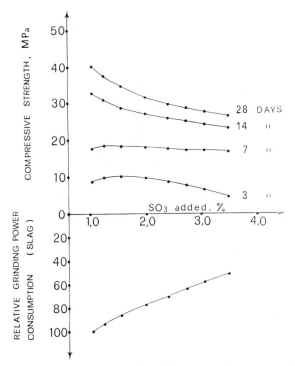

FIG. 4 — *Compressive strength as a function of SO₃ content. Glass content of slag 97%.*

uted finely in the slag and cause a more disordered glass structure with increased reactivity or because these crystals act as hydration nuclei [*32*].

Heat of Hydration

The heat of hydration values found for PBFSCs and its corresponding PCs, all ground to the same Blaine fineness (380 m^2/kg) in a laboratory mill, using 30 different clinkers and always the same slag, do not allow one to form a clear picture of the effect of the clinker's composition on the development of the heat of hydration. One can, however, say from Table 2 that there is a good correlation between the heat developed by PBFSC and the corresponding PC at 3 days, a slight correlation at 7 days, but none at all at 28 days.

Instead, using the same portland clinker and a slag of the same origin and chemical composition, even if with different glass content range 70 to 97%, one notes that there is a strict correlation, even if not a linear one, between the strength and heat of hydration at all ages, as the curve in Fig. 8 shows. The low slope of the second part of the curve shows how it is possible (for example, by increasing the grinding fineness, see Fig. 1) to significantly increase the strengths

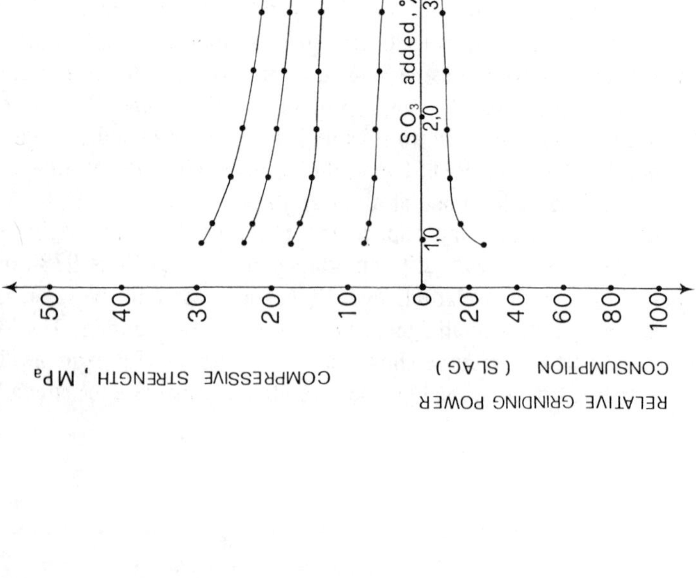

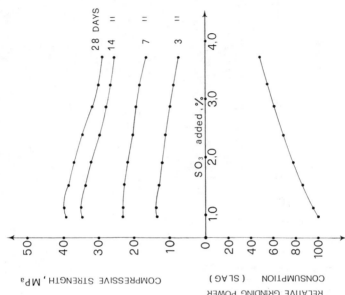

FIG. 6—Compressive strength as a function of SO_3 content. Glass content of slag 51%.

FIG. 5—Compressive strength as a function of SO_3 content. Glass content of slag 51%.

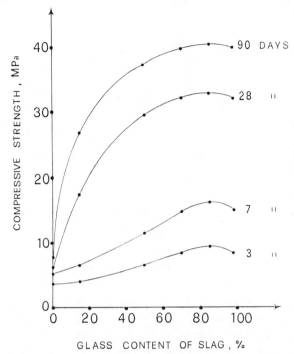

FIG. 7—*Compressive strength as a function of glass content of the slag.*

with slight increases in the heat of hydration. Thus it is possible, for example, to pass from 20 to 40 MPa while heat of hydration increases of only 8 cal/g.

Sulfate Resistance

The results of the tests according to proposed ASTM method and according to Mehta test, obtained with cements made from the same portland clinker ($C_3A = 8\%$; $C_3S = 52\%$) and slag of the same chemical composition and the same origin but with different glass content, are shown in Table 3. The cements

TABLE 2—*Regression lines between heat of hydration of PBFSCs and of corresponding PCs.*

HS (3) = 0.46 HP (3) + 7.8	$r = 0.49$
HS (7) = 0.35 HP (7) + 23.7	$r = 0.32$
HS (28) and HP(28)	$r = 0.01$
$n = 40$	
r scheduled: 99% 0.40	
95% 0.31	

Where—
   HS(3), HS(7), HS(28) are heats of hydration of PBFSC at age of 3, 7, and 28 days, respectively.
   HP (3), HP (7), HP (28) are heats of hydration of the corresponding PC at age of 3, 7, and 28 days, respectively.

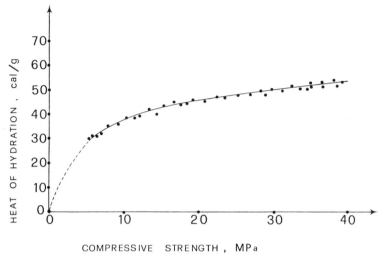

COMPRESSIVE STRENGTH , MPa

FIG. 8—*Heat of hydration as a function of compressive strength.*

had a SO_3 content equal to 1.5% and were ground at the same power consumption. It is seen that slags containing small percentages of crystals do not worsen the resistance to sulfate in mortars made with cement containing them, unlike those with a lower (61%) glass content: this is so because in PBFSC the high resistance to sulfates is due to the impermeability of the mortar to SO_4^- ions [*36–38*], an impermeability which is probably improved by that same mechanism which we have seen ensures the increase in compressive strengths in the presence of small percentages of crystalline slag.

The results showing the effect of the Blaine fineness of blended cement on sulfate resistance of cements prepared with portland clinker having $C_3A = 7\%$,

TABLE 3—*Influence of the glass content of slag on the sulfate resistance.*

	Expansion, %		
	Glass Content of the Slag, %		
Immersion Time, days	61	82	98
EXPANSION ACCORDING TO PROPOSED ASTM METHOD			
28	0.032	0.006	0.007
180	0.090	0.027	0.034
MEHTA TEST			
	Glass Content of the Slag, %		
	61	82	98
Strength after sulfate immersion/ strength before sulfate immersion	0.90	1.05	1.04

TABLE 4—*Influence of the Blaine fineness of blended cement on the sulfate resistance.*

| | Expansion, % | | | |
| | Blaine Fineness, m²/kg | | | |
Immersion Time, days	318	422	578	640
EXPANSION ACCORDING TO PROPOSED ASTM METHOD				
28	0.006	0.006	0.005	0.006
180	0.027	0.026	0.024	0.025
MEHTA TEST				
	Blaine Fineness, m²/kg			
	318	422	578	640
Strength after sulfate immersion/ strength before sulfate immersion	1.08	1.09	1.09	1.08

$C_3S = 56\%$, and slag with a 90% glass content, and gypsum added to give 1.6% SO_3, are reported in Table 4. It is noted that the fineness does not affect the behavior to sulfates of the cement.

The results showing the influence of the SO_3 content on sulfate resistance of the cement (Blaine fineness 422 m²/kg) are reported in Table 5. It is seen that the SO_3 content does not influence the behavior to sulfates of the PBFSC. It is noted that the experiment refers to the SO_3 content range laid down by ASTM Specification for Blended Hydraulic Cements (C 595-82).

Resistance to Alkali-Aggregate Reaction

Influence by the variations in gypsum content of the PBFSC on values of the expansion, due to the alkali-aggregate reaction, is shown in Table 6. The cements

TABLE 5—*Influence of the SO₃ content on the sulfate resistance.*

| | Expansion, % | | |
| | SO₃ Content, % | | |
Immersion Time, days	0.97	1.90	2.95
EXPANSION ACCORDING TO PROPOSED ASTM METHOD			
28	0.006	0.006	0.005
180	0.026	0.028	0.025
MEHTA TEST			
	SO₃ Content, %		
	0.97	1.90	2.95
Strength after sulfate immersion/ strength before sulfate immersion	1.09	1.07	1.07

TABLE 6—*Influence of the SO₃ content on the expansion caused by the alkali-aggregate reaction.*

	Expansion, %	
SO₃ Content, %	14 days	180 days
0.8	0.015	0.022
1.6	0.016	0.023
2.3	0.015	0.023
3.0	0.014	0.024

were prepared by grinding to 380 m^2/kg Blaine portland clinker with 1.2% alkali equivalent and slag, 71% glass content, with 0.4% alkali equivalent. Examination of the values reveals that possible variations in the SO_3 content do not produce any variations of the expansion.

The effect of the grinding fineness is shown in Fig. 9 where values are reported relating to a cement consisting of a portland clinker with a 1.1% alkali equivalent and slag with 0.35% alkali equivalent. It is seen that the alkali-aggregate expansion diminishes with the growth in the fineness of the cement.

With respect to the influence of glass content of the slag, the values of tests carried out on cements, made with slag with different glass content (alkali equivalent content of portland clinker equal to 1.1%) and ground to equal power consumption, are given in the diagram shown in Fig. 10. It is again noted that the presence of small percentages of crystalline fraction, in some way affecting the reactivity of the glass fraction, improve the characteristics of the cements made with it.

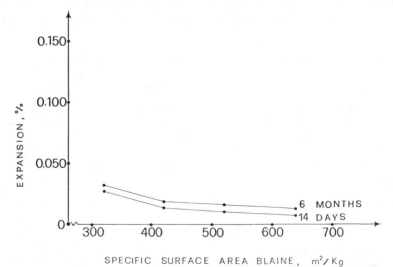

SPECIFIC SURFACE AREA BLAINE , m^2/ Kg

FIG. 9—*Influence of the Blaine fineness on the expansion caused by the alkali-aggregate reaction.*

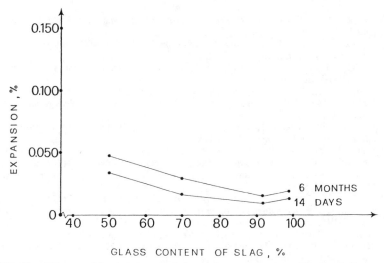

FIG. 10—*Influence of the glass content of slag on the expansion caused by the alkali-aggregate reaction.*

Conclusions

Compressive strength values and heat of hydration of portland blast-furnace slag cements can be affected by parameters outside the cement producer control and essentially attributable to variations in the characteristics of the blast-furnace slags such as glass content and chemical composition. The experimental results show in which way appropriate corrections concerning gypsum content and grinding fineness can positively influence the phenomena induced by the variations in the quality of the blast-furnace slag.

The results further show that resistance to sulfates and alkali-aggregate reaction remain very strong, if the slag content in the cement is high, irrespective of the glass content of the slag as well as of gypsum amount and grinding fineness of of the blended cement.

References

[1] Regourd, M., Hornain, H., and Mortureaux, B., *Revue des Matériaux de Construction,* No. 699, March-April 1976, pp. 83–86.
[2] Daimon, M., *7th International Congress on the Chemistry of Cement,* Paris, 1980, Vol. I, pp. III·2/1–III·2/8.
[3] Locher, F. W., *Zement-Kalk-Gips,* Vol. 66, No. 9, Sept. 1977, pp. 420–429.
[4] Schröder, F., *Proceedings of the Fifth International Symposium on the Chemistry of Cement,* Tokyo 1968, Cement Association Japan, Tokyo 1969, Vol. IV, pp. 149–207.
[5] Smolczyk, H. G., *Revue de Métallurgie,* May 1978, pp. 275–280.
[6] Frigione, G. and Marotta, R., *Giornale del Genio Civile,* Fasc. 7–8–9, July–Aug. 1975, pp. 311–316.
[7] Hogan, F. J. and Mensel, J. W., *Cement, Concrete, and Aggregate,* Vol. 3, No. 1, Summer 1981, pp. 40–52.
[8] Frigione, G. and Marotta, R., *World Cement Technology,* Vol. 12, No. 2, March 1981, pp. 73–78.

[9] Cattaneo, A. and Frigione, G., *Cement, Concrete, and Aggregate*, Vol. 5, No. 1, June 1983, pp. 42–46.

[10] Smolczyk, H. G., *The VI International Congress on the Chemistry of Cement*, Moscow, 1974, Supplementary Paper, Section III, III·2.

[11] Smolczyk, H. G., *7th International Congress on the Chemistry of Cement*, Paris, 1980, Vol. I, pp. III·1/3–III·1/16.

[12] Neville, A. M., *Properties of Concrete*, Pitman Publishing, London, 1975, p. 70.

[13] Keil, F., *Cemento Fabricacion-Propiedades-Aplicaciones*, Editores Técnicos Asociados, Barcelona, 1973, p. 258.

[14] Thomas, A., "Metallurgical and Slag Cement, the Indispensable Energy Savers," IEEE Cement Industry Technical Conference, Innisbrook, FL, 20–24 May 1979.

[15] Venuat, M., *Moniteur des Travaux Publics et du Bâtiment*, Vol. 69, No. 53, 1972, pp. 65–70.

[16] Johansson, S. E., *Silicates Industriels*, Vol. XLIII, No. 7–8, July–Aug., 1978, pp. 139–143.

[17] Totani, Y., Saito, Y., Kageyama, M., and Tanaka, H., *7th International Congress on the Chemistry of Cement*, Paris, 1980, Vol., II, pp. III·95–III·98.

[18] Von Euw, M., *Ciments, Betons, Platres, Chaux*, No. 728, 1/1981, pp. 21–24.

[19] Regourd, M., *7th International Congress on the Chemistry of Cement*, Paris, 1980, Vol. I, pp. III·2/10–III·2/26.

[20] Terrier, P., *Cilam Informations*, Paris, France, No. 8, 1st Trim., 1973, pp. 1–6.

[21] Miyairi, H., Furukawa, R., and Saito, K., *Review of the 29th General Meeting Cement Association Japan*, Tokyo, May 1975, pp. 73–75.

[22] Alsted Nielsen, H. C., *Silicates Industriels*, Vol. XLVIII, No. 4–5, April–May 1983, pp. 87–92.

[23] Venuat, M., *Revue des Matériaux de Construction*, No. 692, Jan.–Feb. 1975, pp. 30–35.

[24] Spellman, L. U., *Concrete International*, Vol. 4, No. 7, July 1982, pp. 66–71.

[25] Frigione, G. and Sersale, R., *Silicates Industriels*, Vol. XLVIII, No. 1, 1983, pp. 23–28.

[26] Frigione, G. and Sersale, R., *Hydraulic Cement Pastes: Their Structure and Properties*, Sheffield, April 1976, pp. 326–329.

[27] Frigione, G., "Gypsum in Cement," in *Advances in Cement Technology*, S. N. Ghosh, Ed., Pergamon Press, Oxford, 1983, pp. 485–535.

[28] Hawthorn, F., Demoulian, E., Gourdin, P., and Vernet, C., *7th International Congress on the Chemistry of Cement*, Paris, 1980, Vol. II, pp. III·145–III·150.

[29] Frigione, G. and Sersale, R., *Cement and Concrete Research*, Vol. 15, 1985, pp. 159–166.

[30] Frigione, G., *Rend. Acad. Scienze Fisiche e Matematiche Società Nazionale Scienze Lettere e Arti in Napoli*, Vol. 51, 1984, pp. 49–60.

[31] Mehta, P. K., *Journal of the American Concrete Institute, Proceedings*, Vol. 72, No. 10, Oct. 1975, pp. 573–575.

[32] Hooton, R. and Emery, J. J., *Fly Ash, Silica Fume, Slag and Other Mineral By-Products in Concrete*, Publication SP-79, American Concrete Institute, Detroit, 1983, Vol. II, pp. 943–954.

[33] Frigione, G. and Di Leva, R., *Cemento*, Vol. 72, No. 1, Jan.–March, 1975, pp. 13–24.

[34] Frigione, G. and Sersale, R., *American Ceramic Society Bulletin*, Vol. 62, No. 11, Nov. 1983, pp. 1275–1279.

[35] Demoulian, E., Gourdin, P., Hawthorn, F., and Vernet, C., *7th International Congress on the Chemistry of Cement*, Paris, 1980, Vol. II, pp. III·89–III·94.

[36] Bakker, R. F. M., *Ciments, Betons, Platres, Chaux*, No. 734, 1/1982, pp. 49–53.

[37] Locher, F. W., *Zement-Kalk-Gips*, Vol. 55, No. 9, Sept. 1966, pp. 395–401.

[38] Frigione, G., *Rend. Acad. Scienze Fisiche e Matematiche Società Nazionale Scienze Lettere e Arti in Napoli*, Vol. 51, 1984, pp. 129–144.

Vladimir S. Dubovoy,[1] Steven H. Gebler,[1] Paul Klieger,[1] and David A. Whiting [1]

Effects of Ground Granulated Blast-Furnace Slags on Some Properties of Pastes, Mortars, and Concretes

REFERENCE: Dubovoy, V. S., Gebler, S. H., Klieger, P., and Whiting, D. A., "**Effects of Ground Granulated Blast-Furnace Slags on Some Properties of Pastes, Mortars, and Concretes,**" *Blended Cements, ASTM STP 897*, G. Frohnsdorff, Ed., American Society for Testing and Materials, Philadelphia, 1986, pp. 29–48.

ABSTRACT: Tests were performed under separate contracts at different times for three manufacturers of ground granulated blast-furnace slag to determine the performance of their products with portland cement. Physical tests were performed on pastes, mortars, and concretes to determine both freshly mixed and hardened properties. The effects of gypsum addition, fineness of slag, and levels of replacement in cement-slag combinations were evaluated. Physical tests performed on pastes and mortars included time of setting, compressive strength, and volume stability. In the mortar compressive strength series, various curing regimens were evaluated with respect to strength development. Included were normal moist curing at 23°C (73°F) and 4.4°C (40°F) and 100% relative humidity, and accelerated atmospheric curing [71°C (160°F)] and 100% relative humidity. Tests conducted on concretes containing slag included compressive strength, freeze-thaw durability, and resistance to deicer chemicals (scaling). Measurements of air-void systems of hardened concretes were made. Results generally show that use of slag can be beneficial without resulting in significant technical problems or adverse construction problems.

KEY WORDS: accelerated curing, air entrainment, blast furnace slag, blended cements, cold weather construction, compressive strength, concrete durability, curing, drying shrinkage, durability, fineness, freeze-thaw durability, granulation, gypsum, heat of hydration, mineral admixtures, scaling, setting, slag

This paper summarizes the results of three separate laboratory studies directed towards the evaluation of various ground granulated blast-furnace slags (GGBFSs) and slag-cement combinations as to their influence on physical prop-

[1]Research engineer, senior research engineer, consultant, and senior research engineer, respectively, Concrete Materials/Technical Services Department, Construction Technology Laboratories, a Division of the Portland Cement Association, Skokie, IL 60077-4321.

erties of pastes, mortars, and concretes. These contract research studies were conducted at the Construction Technology Laboratories, during the period between 1978 and 1982.

Blast-furnace slag (BFS), the residual by-product of iron and steel manufacture has been used as a supplementary cementitious material. Ground blast-furnace slag (GBFS) can be incorporated at either of two points in the production of concrete: as an ingredient mixed with portland cement (PC) at the cement plant to produce a blended cement or as a mineral admixture employed in conjunction with PC at the concrete batch plant.

Applications of GGBFS have been relatively well understood and employed in varying degrees for decades both in this country and abroad. However, chemical interactions are complex and require extensive research and development to advance the level of understanding of BFS uses. A major reason for this complexity is the generally heterogeneous composition and properties of this type of residual material; this factor makes the generic performance of slag in concrete less predictable than that of PC alone.

This paper summarizes some of the physical properties of several slags used in these studies, as well as physical properties of pastes, mortars, and concretes prepared with these slags-cement combinations.

Findings and Conclusions

1. For both mortars and concretes, an optimum level of slag replacement exists for which strength is maximized. Generally, this level is approximately 50% of cement weight replacement.

2. At normal temperatures, early age strength development is retarded when slags are used. The point in time at which the strength of a slag-cement mixture becomes equivalent to that of a straight cement mixture is a function of the particular slag-cement combination being used.

3. Accelerated curing increases early age strength development for slag-cement mixtures.

4. At low temperature, replacement of cement with slag in mortars results in a substantial loss of strength through 7 days, the latest test age in these studies.

5. In general, strength of cement-slag mixtures increases with an increase in slag fineness.

6. Setting time of pastes is retarded when a portion of the cement is replaced by slag.

7. Durability of air-entrained slag-cement concretes, with regard to freezing and thawing in water, is essentially equivalent to that of concretes containing solely portland cement.

8. Air-entrained slag-cement concretes are somewhat less resistant to laboratory deicer scaling tests than air-entrained concrete containing solely PC, despite the fact that both concretes had adequate air-void systems.

TABLE 1—*Chemical properties of slag.*[a]

Chemical Analysis, %	Slag		
	A	B[b]	C[c]
SiO_2	36.18	34.20	33.80
Al_2O_3	10.20	13.80	11.70
Fe_2O_3	0.60	...[d]	1.35
CaO	39.85	42.80	30.00
MgO	11.22	6.46	16.00
SO_3	0.29	0.06[c]	2.84
S	0.48	...	...
Na_2O	0.21	0.20[c]	0.23
K_2O	0.37	0.38[c]	0.56
Total alkalies as Na_2O	0.45	0.45[c]	0.60
Glass content	...	98.80	...
Loss on ignition	0.12	...	1.45
Free CaO	0.06	...	...
Insoluble residue	0.36	...	...

[a]As-received slag composition. Chemical analysis furnished by sponsor of research except where noted.
[b]Slag produced outside United States of America.
[c]Chemical analyses conducted by Construction Technology Laboratories.
[d]Test results not available.

Materials

Three GBFSs from different sources were ground and denoted A, B, and C. Various Type I PCs were used in combination with these slags. It should be noted that some of the data are not directly comparable, as in most instances different cements were used with each slag. Chemical properties of the slags and cements are presented in Tables 1 and 2.

Test Program

Slag-cement paste, mortar, and concrete physical properties were evaluated. Test procedures are shown next. Each method followed the prescribed ASTM procedures except for mix proportions and some test ages.

Part I-Paste and Mortar Tests

(*a*) Time of setting, ASTM Test for Time of Setting of Hydraulic Cement by Vicat Needle (C 191-82), (*b*) Heat of hydration, ASTM Test for Heat of Hydration of Hydraulic Cement (C 186-82), (*c*) Compressive strength, ASTM Test for Compressive Strength of Hydraulic Cement Mortars (C 109-80), (*d*) Drying shrinkage, ASTM Test for Length Changes of Hardened Cement Mortar and Concrete (C 157-80), and (*e*) Effects of various curing procedures on strength.

TABLE 2 — *Chemical properties of cement.*[a]

Chemical Analysis, %	Cement		
	A	B	C[b]
SiO_2	21.58	22.34	20.77
Al_2O_3	4.66	4.18	5.46
Fe_2O_3	3.50	3.53	2.21
CaO	63.04	64.57	65.12
MgO	3.59	1.16	1.05
SO_3	2.58	2.46	2.44
Na_2O	0.25	0.35[(2)]	... [c]
K_2O	0.39	0.30[(2)]	...
Total alkalies as Na_2O	0.51	0.55[(2)]	0.34
Loss on ignition	0.77	...	1.44
Free CaO	0.12	...	...
Insoluble residue	0.34	...	1.44
C_3S	48.80	52.86	60.00
C_2S	25.10	24.21	14.00
C_3A	6.40	5.11	11.00
C_4AF	10.70	10.76	6.00

[a]Chemical analyses furnished by sponsor of research except where noted.
[b]Chemical analyses conducted by Construction Technology Laboratories.
[c]Test results not available.

Part-II Concrete Tests

(*a*) Compressive strength, ASTM Test for Compressive Strength of Cylindrical Concrete Specimens (C 39-83a), (*b*) Resistance to deicer scaling, ASTM Test for Scaling Resistance of Concrete Surfaces Exposed to Deicing Chemicals (C 672-76), and (*c*) Freeze-thaw durability, ASTM Test for Resistance of Concrete to Rapid Freezing and Thawing (C 666-80), Procedure A.

Discussion of Test Results

Paste and Mortar Tests

Setting time — Setting times were determined for various replacement levels of GBFS B and C. Results, summarized in Table 3, indicate that the use of 50% slag for cement significantly retards the setting time compared to 25% or even 40% slag replacement. Initial set retardations for slag B were 32 min and 1 h:26 min for the 25 and 50% slag content mixtures, respectively. Final set retardations were 45 min and 1 h:15 min for the 25 and 50% slag content mixtures, respectively. Mixes with slag C exhibited 30 min initial set and 30 min final set retardation for 40% slag content mix.

Heat of Hydration — A series of heat of hydration tests was run for slag A and C pastes, varying slag content and, in the case of slag A, Blaine fineness and sulfur trioxide (SO_3) content. Results, presented in Table 4, indicate that addition of slag A yielded a higher heat of hydration, while addition of slag C yielded a

TABLE 3—*Setting time. Slags B and C.*

Cement-Slag Combination		Setting Time, h:min		Retardation h:min	
Slag B	Slag C	Initial	Final	Initial	Final
100% cement	. . .	2:16	3:00	. . .	. . .
25% slag, 75% cement	. . .	2:48	3:45	0:32	0:45
50% slag, 50% cement	. . .	3:42	4:15	1:26	1:15
. . .	100% cement	2:25	3:00	. . .	. . .
. . .	40% slag, 60% cement	2:55	3:30	0:30	0:30

lower heat of hydration as compared with the control (no slag) pastes. These differences in heat of hydration correspond to the relative strength influence produced by these two slags in mortars and concretes.

Table 4 indicates that for slag A, as the SO_3 content increases, the heat of hydration increases. Also, as expected, the coarser ground slag A (377 m^2/kg) resulted in considerably lower heat of hydration at 7 days. At a SO_3 level of 3.00%, the heat of hydration increased when slag A was added to the mixture. The heat of hydration was affected only marginally when the slag A content was increased from 40 to 65%. Fineness and SO_3 content obviously play an important role in determining the amount of heat generated.

Mortar Compressive Strength —The slag-cement mortar cube compressive strengths were examined from the standpoint of replacement level and slag fineness. Figures 1 and 2 and Table 5 show the strength development curves for slags A and B at various replacement levels. The data for continuous moist curing

TABLE 4—*Heat of hydration. Slags A and C.*

Source and Amount of Slag	Slag Blaine Fineness, m^2/kg	SO_3 Content,[a] %	Heat of Hydration, cal/g	
			7 days	28 days
		CEMENT A		
A–0%	. . .	2.58	80.4	88.6
A–0%	. . .	3.00[b]	82.3	91.7
A–40%	559	1.66	90.0	96.6
A–40%	377	1.66	84.1	97.1
A–40%	377	3.00[b]	97.4	104.7
A–65%	559	1.09	89.3	98.9
A–65%	377	1.09	78.9	89.8
A–65%	377	3.00[b]	96.9	102.2
		CEMENT C		
C–0%	. . .	2.44	84.6	98.0
C–40%	453	2.60	76.5	96.2

[a]Total sulfate content of cement-slag mixture.
[b]Sulfate content modified by blending of extra gypsum to a cement-slag mixture.

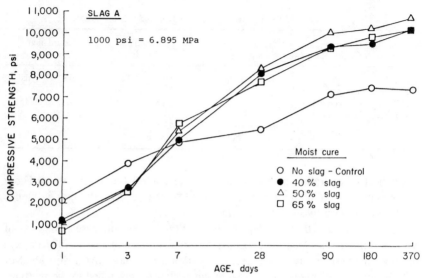

FIG. 1—*Mortar cube compressive strength development for various slag replacements. Mixed at 23°C (73°F) and cured at 23°C (73°F) and 100% RH.*

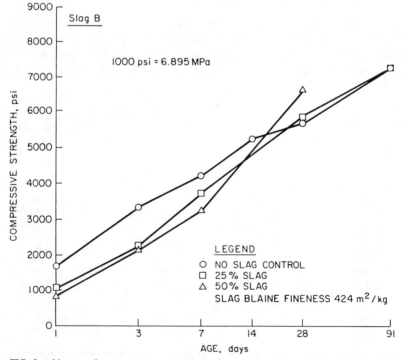

FIG. 2—*Mortar cube compressive strength development for various slag replacements. Mixed at 23°C (73°F) and cured at 23°C (73°F) and 100% RH.*

TABLE 5—*Compressive strength of mortar and slag replacement level. Slags A and B.*

Cure	Source and Amount of Slag	Compressive Strength, psi[a]							
		1 day	3 days	7 days	14 days	28 days	90 days	180 days	365 days
	CEMENT A								
Moist cure @23°C (73°F)	A–0%	2100	3840	4860	...[b]	5410	7050	7380	7280
	A–40%	1260	2720	4920	...	8060	9350	9500	10 140
	A–50%	1080	2720	5350	...	8360	9940	10 240	10 620
	A–65%	720	2540	5740	...	7640	9310	9790	10 150
	CEMENT B								
	B–0%	1700	3330	4380	5260	5820	7180	...	...
	B–25%	1050	2370	3710	4680	5850	7180	...	...
	B–50%	860	2150	3250	...	6640	...	...	...
	CEMENT A								
Accelerated cure @71.1°C (160°F)	A–0%	4950	...	5440	...	6850	...	...	...
	A–40%	5480	...	6360	...	7820	...	...	...
	A–50%	5810	...	6680	...	6700	...	...	...
	A–65%	5990	...	6700	...	7710	...	...	...
Mix and cure @4.4°C (40°F)	A–0%	160	1350	3060	...	...	...	...	...
	A–40%	90	600	1520	...	...	...	...	...
	A–50%	80	440	1180	...	...	...	...	...
	A–65%	70	260	920	...	...	...	...	...

[a]To convert pounds per square inch to megapascals, multiply by 6.895×10^{-3}.
[b]Tests not run.

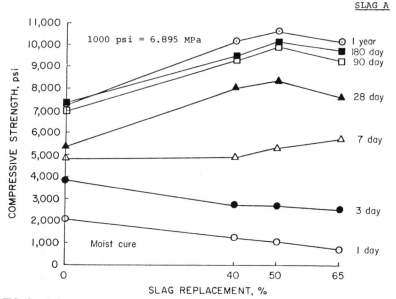

FIG. 3—*Influence of slag replacement on mortar cube compressive strength. Mixed at 23°C (73°F) and cured at 23°C (73°F) and 100% RH.*

at 23°C (73°F) indicate that both types of slag-cement mixtures develop considerably less compressive strength at early ages (1 to 3 days) than the control, over the range of replacement levels employed. The strengths of mixtures made with slag A equaled the control strength at about 6 days, while the strength of mixtures made with slag B did not reach the control strength until about 25 days.

Figure 3 shows that at ages greater than 28 days there seems to be an optimum level of slag which maximizes strength. For slags A and B this replacement level is about 50% of cement.

The effect of slag fineness on compressive strength is shown in Fig. 4 and Table 6 for slag B and in Fig. 5 and Table 6 for slag C. Slag B was ground to three Blaine fineness levels (424, 501, 595 m²/kg) whereas slag C was ground to six Blaine fineness levels (451, 479, 509, 548, 576, and 600 m²/kg). Test data indicate that mortars containing finer GBFS produced somewhat higher compressive strengths than ones with coarser ground slag, as one would expect. As shown in Fig. 6 and Table 6, the compressive strength of the slag-cement mixtures with slag B ground to a Blaine fineness of 595 m²/kg overtook the no-slag mixture at 7 days. The slag ground to 501 m²/kg Blaine fineness overtook the no-slag mixture at 14 days. The slag ground to 424 m²/kg Blaine fineness overtook the no-slag mixture at 28 days. For slag C, however, as indicated in Fig. 7 and Table 6, only those ground to Blaine fineness 509 m²/kg and finer overtook control (no slag) mixes at 28 days. Therefore, the effect of slag fineness on

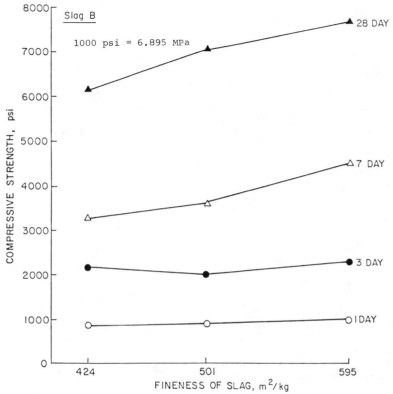

FIG. 4—*Influence of fineness of slag on mortar cube compressive strength. Mixed at 23°C (73°F) and cured at 23°C (73°F) and 100% RH.*

strength, although generally consistent, depends to a large extent on the particular slag used. Note that at early ages, as indicated in Figs. 6 and 7, the compressive strength is not as dependent on slag fineness as it is at later ages.

Drying Shrinkage — Results of drying shrinkage tests on mortar bars are shown in Table 7. Results indicate that addition of these ground blast-furnace slags increased drying shrinkage. For example, drying shrinkages of slags A and B mixtures were significantly greater than those of their control (no slag) counterparts. At 16 weeks of drying the differences were 34 and 25% for slags A and B mixtures, respectively. Other data shown in Table 7 indicate that the addition of gypsum did not have a consistent effect on drying shrinkage. Table 7 shows that for mixtures with slag A, the shrinkage was somewhat less for an SO$_3$ content of 3.00% as compared with an SO$_3$ content of 1.44%. For mixtures with slag B, however, shrinkage with 3.30% SO$_3$ content was slightly greater than that with 2.01% SO$_3$.

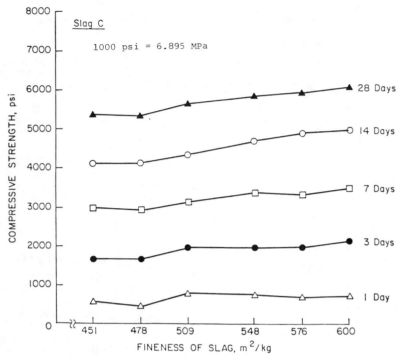

FIG. 5—*Influence of fineness of slag on mortar cube compressive strength. Mixed at 23°C (73°F) and cured at 23°C (73°F) and 100% RH.*

Effect of Various Curing Procedures on Strength Development — Mortar cubes with slag A were subjected to accelerated curing and low temperature curing and tested for compressive strength. The accelerated cure cycle consisted of 5 h preset, followed by 4 h temperature rise at the rate of 14°C/h (25°F/h), followed by 16 h at 71°C (169°F) and subsequent moist storage at 23°C (73°F). The continuous low temperature curing was at 4.4°C (40°F) and 100% relative humidity.

Figures 8 and 9 and Table 5 indicate that mixtures containing slag that were accelerated cured showed improved compressive strength at ages of 1, 7, and 28 days relative to the control (no slag). Accelerated curing had a beneficial effect on strength at all ages for all slag replacement levels. There seems to be no significant difference in strength between slag replacement levels of 50 and 65% at 1 and 7 days. The 28-day strength development curve shown in Fig. 8 suggests that there is an optimum slag level which produces the maximum accelerated-cured 28-day strength (about 50% slag replacement of cement).

Figures 10 and 11 and Table 5 show decreasing strengths with increasing levels of slag replacement at mixing and curing temperatures of 4.4°C (40°F). When mortars are mixed and cured at low temperature, compressive strength develop-

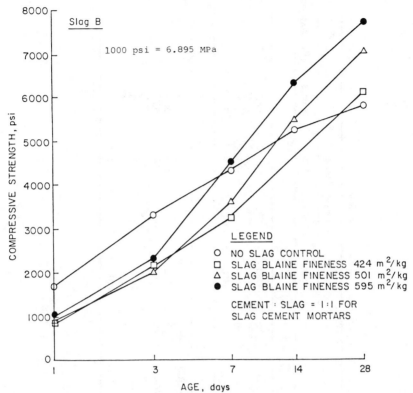

FIG. 6—*Mortar cube compressive strength development for various slag finenesses. Mixed at 23°C (73°F) and cured at 23°C (73°F) and 100% RH.*

ment at 1, 3, and 7 days is inversely proportional to increase in slag content. The use of slag in cold weather concreting should be evaluated on a job-by-job basis.

Concrete Tests

Concrete Compressive Strength —Compressive strength determination was performed on moist-cured concretes made with slags A and C. Concretes contained 19.0 mm (¾-in.) maximum size gravel and a concrete sand having a fineness modulus ranging between 2.7 and 2.9. Table 8 shows compressive strength development of concretes prepared with a water-cementitious ratio of 0.55, with slag A replacements of 40%, 50%, and 65% and slag C replacement of 40%.

Compressive strength of concrete with slag A overtook the control at an age of about 7 days and by 28 days showed 3 to 13% higher compressive strength than the concrete without slag, depending on replacement level. The higher the

TABLE 6—*Compressive strength of mortar and slag Blaine fineness. Slags B and C.*

Source and Amount of Slag	Slag Blaine Fineness, m^2/kg	W/C[a]	Compressive Strength, psi[b]					
			1 day	3 days	7 days	14 days	28 days	91 days
CEMENT B								
B–0%	...	0.48	1700	3300	4380	5260	5820	7180
B–50%	424	0.46	860	2150	3250	...[c]	6640	...
B–50%	501	0.47	890	2020	3630	5520	7070	...
B–50%	595	0.46	980	2320	4530	6350	7730	...
CEMENT C								
C–0%	...	0.49	1980	3330	4330	...	5530	...
C–50%	451	0.48	670	1770	2950	4200	5340	...
C–50%	479	0.48	640	1760	2900	4180	5330	...
C–50%	509	0.48	690	1930	3120	4380	5700	...
C–50%	548	0.48	680	1940	3350	4730	5810	...
C–50%	576	0.48	660	1990	3320	4860	5920	...
C–50%	600	0.48	670	2040	3460	4920	6050	...

[a]Water-cement ratio includes cement and slag.
[b]To convert pounds per square inch to megapascals, multiply by 6.895 × 10^{-3}.
[c]Tests not run.

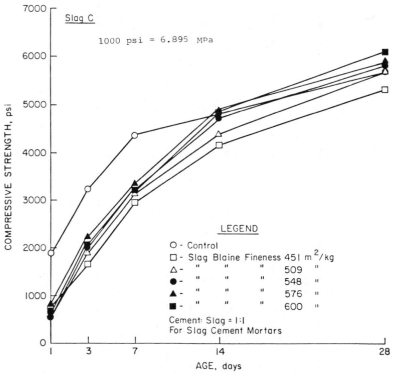

FIG. 7—*Mortar cube compressive strength development for various slag finenesses. Mixed at 23°C (73°F) and cured at 23°C (73°F) and 100% RH.*

slag content, the lower the strength gain. Concrete with slag C showed greater compressive strength by 91 days.

In addition, as in the mortar series, there is apparently an optimum level of slag which maximizes the concrete compressive strength. For instance, for slag A this level was found to be about 40%. At early ages (1 and 3 days), both slag mixtures exhibited lower compressive strength than the control mixtures.

Adjustments in the chemical composition of the slag may aid in early strength development. For example, Warris [1] has shown that increasing the alumina level (from 10 to 16%) and reducing the glass content (from 98 to 79%) may increase strength by more than 33% at 1, 3, and 7 days.

Resistance to Deicer Scaling —Deicer scaling resistance was determined on air-entrained concrete with slag A and concrete without slag. Results are shown in Table 9. Test and reference slabs were moist cured for 14 days followed by storage in laboratory air at 23°C (73°F) and 50% relative humidity for 14 days. Also presented in Table 9 are the air-void parameters for test and reference concretes.

TABLE 7—Drying shrinkage of mortar. Slags A and B.

Source and Amount of Slag	SO$_3$[a] Content, %	W/C[b]	Drying Shrinkage, %								
			1 day	4 days	7 days	14 days	28 days	8 weeks	16 weeks	32 weeks	64 weeks
			CEMENT A								
A–0%	2.58	0.50	...	...	...	0.044	0.072	0.088	0.100	0.106	0.113
A–0%	3.00[c]	0.50	...	...	...	0.043	0.070	0.085	0.099	0.107	0.113
A–50%	1.44	0.50	...	...	...	0.083	0.121	0.147	0.161	0.171	0.179
A–50%	3.00[c]	0.50	...	...	...	0.037	0.066	0.099	0.126	0.141	0.158
			CEMENT B								
B–0%	2.46	0.49	0.020	0.041	0.053	0.073	0.082	0.090	0.098	0.104	0.113
B–50%	1.26	0.49	0.039	0.062	0.077	0.094	0.109	0.123	0.129	0.132	0.150
B–50%	2.01[c]	0.49	0.034	0.052	0.071	0.088	0.110	0.119	0.128	0.137	0.156
B–50%	2.63[c]	0.49	0.034	0.057	0.071	0.091	0.115	0.128	0.138	0.148	0.173
B–50%	3.30[c]	0.49	0.029	0.051	0.068	0.087	0.111	0.127	0.138	0.148	0.174

[a]Total sulfate content of slag-cement mixture.
[b]Water-cement ratio includes cement and slag.
[c]Sulfate content modified by blending of extra gypsum to a cement-slag mixture.

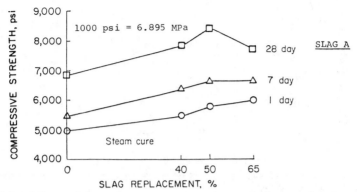

FIG. 8—*Influence of slag replacement on mortar cube compressive strength and accelerated curing.*

Concrete slabs with slag A and their control counterparts were subjected to freezing and thawing at a rate of one full cycle per day. Flake calcium chloride was applied to the ice on the slab surface at the beginning of each thawing cycle. After thawing was completed, the slabs were washed and fresh water replaced for freezing. After 300 cycles, the 50% slag mixture exhibited moderate to severe scaling, and the control slabs exhibited moderate scaling. The data presented in Table 9, with regard to air-void parameters, show the concrete with slag had a better air-void system than the reference concrete. Air-entrained slag-cement concrete is somewhat less resistant to laboratory deicer scaling tests than air-entrained concrete containing solely PC, despite the fact that both concretes had adequate air-void systems. These data are consistent with observations by Klieger and Isberner [2].

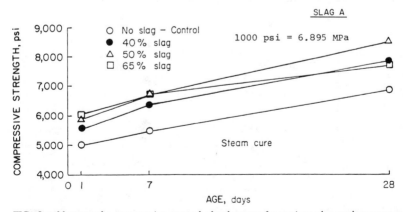

FIG. 9—*Mortar cube compressive strength development for various slag replacements and accelerated curing.*

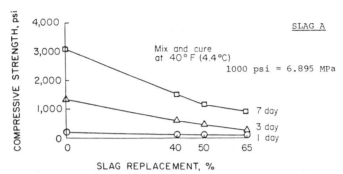

FIG. 10 — *Influence of slag replacement on mortar cube compressive strength and low-temperature mixing and curing.*

Freeze-Thaw Resistance — Freeze-thaw specimens were cast from air-entrained concrete mixtures with 50% slag replacement of cement and a control concrete (no slag). The prisms were moist cured for 14 days and subsequently placed in a freeze-thaw chamber. Length, weight, and change in fundamental transverse frequency were monitored and results are shown in Table 10. While there appears to be a difference in the durability factors between the control and the slag-cement concrete mixtures (98 versus 89, respectively), the relative durability factor of the slag-cement concrete was 91. The durability factors for the slag-cement concrete indicate sound quality concrete. Both concretes exhibited insignificant weight losses and expansions.

Summary and Recommendations

The physical tests conducted on various ground granulated blast-furnace slags indicate that slag as an admixture or used as a blended cement can yield satisfactory concretes with respect to strength development and durability. Since the data show differences in performance between slags, it is recommended that tests

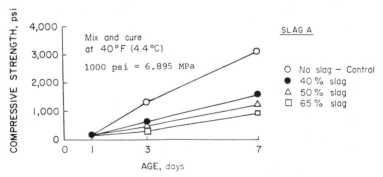

FIG. 11 — *Mortar cube compressive strength development for various slag replacements for low-temperature mixing and curing.*

TABLE 8—*Compressive strength of concrete.[a] Slags A and C.*

Source and Amount of Slag	W/C[b]	Compressive Strength, psi[c]					
		1 day	3 days	7 days	14 days	28 days	91 days
CEMENT A							
A–0%	0.55	...	4680	5800	...[d]	7240	...
A–40%[e]	0.55	...	3930	6130	...	8220	...
A–50%[e]	0.55	...	3870	5670	...	8060	...
A–65%[e]	0.55	...	3710	5700	...	7470	...
CEMENT C							
C–0%	0.56	1230	2770	4110	4840	5890	6300
C–40%f	0.64	600	1790	3400	4570	5400	6820

[a]Mixed at 23°C (73°F) and moist cured.
[b]W/C denotes water-cementitious ratio which includes cement and slag.
[c]To convert from pounds per square inch to megapascals, multiply by 6.895×10^{-3}.
[d]Tests not run.
[e]Slag Blaine fineness = 559 m^2/kg.
fSlag Blaine fineness = 453 m^2/kg.

TABLE 9 — *Air content and deicer scaling.*[a] *Slag A.*

Concretes	Air Content, %		Voids per in., n^b	Specific Surface, α, in.^{2}/in.$^{3\,c}$	Spacing Factor, $\overline{L}$, in.^{d}	Scale rating[e] at Deicer Cycle		
	Fresh	Hardened				100	200	300
Cement (A), 100%, W/C = 0.56,[f] slump, 90 mm, (3.5 in.) cement content, 278 kg/m³ (468 lb/yd³)	4.7	4.2	6.7	639	0.008	2+	3	3
50% slag (A) and 50% cement (A) W/C = 0.53,[f] slump = 95 mm, (3¾ in.), cement content, 139 kg/m³ (234 lb/yd³)	5.7	4.7	9.0	767	0.006	3+	4−	4

[a]Results are the average for three 75 by 150 by 381-mm (3 by 6 by 15-in.) slabs.
[b]To convert from voids inch to voids per millimetre, multiply by 0.0394.
[c]To convert from square inch per cubic inch to square millimetre per cubic millimetre, multiply by 0.0394.
[d]To convert from inches to millimetres, multiply by 25.4.
[e]Where 0 = no scaling,
  1 = slight scaling,
  2 = slight to moderate scaling,
  3 = moderate scaling,
  4 = moderate to severe scaling, and
  5 = severe scaling.
[f]W/C denotes water-cementitious ratio which includes slag and cement.

TABLE 10—*Freeze-thaw durability. Slag A.*

Mix	W/C[a]	Freshly Mixed Air Content, %	Slump, in.[b]	Water Content, lb/yd$^{3\,c}$	Cement and Slag Content, lb/yd$^{3\,c}$	Unit Weight, lb/ft$^{3\,d}$	Changes in Property at 301 Cycles of Freezing and Thawing[e]		
							Expansion, %	Weight, %	Durability Factor
Cement (A), 100%	0.56	5.5	3¼	253	454	146.0	+0.010	−3.25	98
Slag (A), 50%	0.56	6.0	3¼	252	453	145.4	+0.026	−2.42	89
Cement (A), 50%									relative durability factor = 91

[a] W/C denotes water-cementitious ratio which includes slag and cement.
[b] To convert from inches to millimetre, multiply by 25.4.
[c] To convert from pounds per cubic yard to kilograms per cubic metre, multiply by 0.594.
[d] To convert from pounds per cubic foot to kilograms per cubic metre, multiply by 16.019.
[e] Results are the average for three 3 by 3 by 11¼-in. (75 by 75 by 286-mm) prisms.

be conducted on individual slag and cement mixtures to establish and characterize slag performance.

Acknowledgments

The authors would like to thank the Atlantic Cement Company, The Standard Slag Company, and The Warner Slag Company who provided the slags and sponsored this study.

References

[1] Warris, B., *Granulated Blast Furnace Slag for Blended Cement,* Cementa AB, Stockholm, Sweden, page 6.2.
[2] Klieger, P. and Isberner, A., "Laboratory Studies of Blended Cements–Portland Blast-Furnace Slag Cements," *Journal of the PCA Research and Development Laboratories,* Vol. 9, No. 3, 2–22, Sept. 1967.

Ronald H. Mills [1]

Chemical Shrinkage and Differential Sorptions in Mixtures of Portland Cement and Blast-Furnace Slag

REFERENCE: Mills, R. H., "**Chemical Shrinkage and Differential Sorptions in Mixtures of Portland Cement and Blast-Furnace Slag**," *Blended Cements, ASTM STP 897,* G. Frohnsdorff, Ed., American Society for Testing and Materials, Philadelphia, 1986, pp. 49–61

ABSTRACT: The relationship between chemical shrinkage, expressed as volume per unit mass of anhydrous cementing material, and chemically bound water varies over a narrow range with the ratio of portland cement (PC) to blast-furnace slag (BFS). Chemical shrinkage is easily measured and affords an easily executed nondestructive measure of quality for cements made with mixtures of PC and BFS.

The difference between the pore space occupied by water and by kerosene expressed as a fraction of water-filled space is a measure of the volume proportion of hydrate material and is related to relative shrinkage and creep.

KEY WORDS: blast-furnace slag, cement, portland cement, chemical shrinkage, shrinkage, creep, differential sorption

Nomenclature

A	Constant in equation $\sigma = AX^n$ (Mpa)
C	Mass of portland cement (g or kg)
C_1, C_2, etc.	Different portland cements
d	Density (g/mL)
k	Ratio: mass of nonevaporable water to mass of water taken up to maintain saturation (dimensionless g/g)
H	Hydraulicity index $= (\sigma_z - \sigma_i)/\sigma_p - \sigma_i$ (nondimensional)
m	$(V_m - V_L)/V_w$ (dimensionless mass ratio g/g)
n	Exponent in equation $\sigma = AX^n$
r	Ratio: mass of blast-furnace slag to mass of portland cement + blast-furnace slag (dimensionless g/g)

[1]Professor, Department of Civil Engineering, University of Toronto, Ont., Canada.

P	Mass of portland cement in the mixture
R	(Coefficient of determination)$^{1/2}$
S_1, S_2, etc.	Different blast-furnace slags
S	Mass of blast-furnace slag in the mixture
V_w	Volume of water per unit volume of cement paste, mortar, or concrete (dimensionless g/g)
V_L	Volume of liquid other than water, per unit volume, which is taken up by previously dried cement paste, mortar, or concrete (dimensionless mL/mL)
V_h	Volume of hydrate products per unit volume of cement paste (mL/g)
V_k	Volume of kerosene per unit volume of cement paste, mortar, or concrete (dimensionless mL/mL)
W	Mass of mixing water per unit volume of cement paste, mortar, or concrete (kg/m^3 or g/L)
ΔW	Mass of water added to maintain saturation of hydrating cement per unit volume of cement paste, mortar, or concrete (kg/m^3 or g/L)
ΔW^0	Terminal value of ΔW at full hydration
w_o	Water/cement ratio (g/g)
w_n	Mass ratio of nonevaporable water to anhydrous cement (g/g)
w_n^0	Mass ratio of nonevaporable water to unit mass of completely hydrated cement (g/g)
Δw	Mass ratio of water added to maintain saturation to unit mass of anhydrous cement $= \Delta W/C$ (g/g)
α	Mass proportion of cement that has become fully hydrated — w_n/w_n^0 (g/g)
ε	Strain (microstrain)
ε_s	Shrinkage strain (microstrain)
ε_c	Creep strain (microstrain)
σ	Stress (MPa)

Chemical Shrinkage

When water reacts with cement, the new solid formed has a specific volume about 1.6 times that of the anhydrous cement [1, 2] but only about 86% of the starting volume of cement and water [3–5]. Also the gross volume of the porous cement paste increases by about 0.6% [6]. Strength, imperviousness to water and gas, and other desirable properties depend on the extent to which space between cement grains in the fresh concrete is filled by the products of hydration [7, 8].

The volume contraction accompanying hydration, which occurs while the material is saturated, has come to be known as chemical shrinkage to distinguish it from drying shrinkage. Several authors [9–12] have reported a linear relationship between nonevaporable water (w_n) and chemical shrinkage (Δw). Chemical shrinkage is generally measured by weighing the added water necessary

TABLE 1 — *Oxide compositions and physical properties of PCs and BFSs.*

Oxide	PC		BFS			
	C_1	C_2	S_1	S_2	S_3	S_4
SiO_2, %	22.6	22.2	32.9	32.0	30.0	35.6
Al_2O_3 (+TiO_2 + Mn_2O_3), %	5.2	5.3	16.1	19.9	18.6	9.9
Fe_2O_3, %	2.6	2.5	0.7	0.7	0.7	0.4
CaO, %	63.9	65.3	30.6	41.0	39.9	44.0
MgO, %	2.5	1.8	20.0	2.1	5.9	2.9
SO_3, %	1.8	1.8	. . .	1.2	1.0	1.0
Blaine surface area, m^2/kg	300	300	375	361	374	327
Density, g/mL	3.22	3.22	2.96	2.90	2.90	2.90

to maintain saturation [2,3,5]. The widely quoted [7] power curve relating strength and gel-space ratio suggests that strength, which is a universally accepted quality index, may be directly related to chemical shrinkage defined as the ratio of the mass of water taken up by the cement paste in order to maintain saturation to the mass of cement. Although uptake of water is caused by reduction of specific volume of the products of hydration, because it is observed as a weight change, its units are dimensionless gram/gram (g/g) of anhydrous cement.

If such a functional relationship could be established, it would offer attractive possibilities for continuous nondestructive monitoring of concrete quality in the field and for easy evaluation of cementing properties from small samples of clinker and blast-furnace slag (BFS). The prospects for easy evaluation of portland cement (PC) and BFS mixtures provided the stimulus for the work presented herein.

Experimental

Four different BFSs and two different PCs having the oxide compositions and physical properties listed in Table 1 were investigated.

Bottle Hydrated Slurries

Various mixtures of PC and BFS were ball-milled in ½-L pyknometer bottles. 12-mm-diameter ceramic balls provided the grinding action. The starting water:cement (W/C) ratios by mass were over 4. Chemical shrinkage was determined by the mass increment of water, ΔW, needed to keep the pyknometer bottles filled to the mark. Each bottle contained about 116-g cement, and the total uptake of water, ΔW^0, at full hydration was typically about 7.6 g. This value was determined by weighing, with an average standard deviation of 0.3 g for 16 groups of three replicates. The ΔW measurements were terminated when the incremental mass reduced to about 1 mg/g of cement per week. This generally took place at about 2 months age. The mass change per unit of cement $\Delta w = \Delta W/C$, where C was the mass of anhydrous cement in the pyknometer

TABLE 2—*Constants obtained by hydration of PC C_1 and BFS S_1 in pyknometer bottles.*

d for Combined Material, g/mL	Average Values for Full Hydration				
	r	w_n^0	V_h	Δw	$V_h/\Delta w$
3.22	1.00	0.253	0.398	0.0654	0.135
3.16	0.75	0.225	0.396	0.0746	0.117
3.09	0.50	0.261	0.397	0.0842	0.101
3.03	0.25	0.264	0.393	0.0974	0.085
2.99	0.10	0.239	0.380	0.1021	0.081

Regression equation $V_h/\Delta w = 0.07 + 0.06r$ ($R^2 = 0.99$),
  d = density, g/mL,
  r = ratio $[C_1/(C_1 + S_1)]$,
  w_n^0 = mass of nonevaporable water per unit mass of cement, g/g,
  V_h = specific volume of hydration products, mL/g,
  Δw = chemical shrinkage measured as mass of water absorbed per unit mass of cement,
      g/g, and
  $V_h/\Delta w$ = ratio of volume of hydration products per unit of chemical shrinkage, mL/g.

NOTE—The value of w_n^0 is close to published values for PC (for example Refs 2 and 27). In the case of mixtures of PC and BFS the values correspond to an end point where the rate of increase in w_n was negligible as illustrated in Ref 8.

bottle. This mass change was caused by chemical shrinkage but was not exactly equal to it since the specific volume of pore water ≠ 1.

The mass of nonevaporable water, w_n, per gram of cement corresponding to various Δw was determined by first drying slurry samples at 110°C, then igniting them at 1050°C. These determinations were continued for about 3 years.

The results of these tests are summarized in Table 2. The values for the density, d, the mass of nonevaporable water per gram of cement at ultimate hydration, w_n^0, and the ratio of $w_n/\Delta w = k$ facilitated the calculation of porosity, degree of hydration, α, and volume concentration, X, of hydration product in the available

TABLE 3—*Mixture proportions for solid specimens cast as cubes and in pyknometers for Δw measurements (units are g/L).*

Mixture	PC, C_1	BFS, S_1	Quartzite Aggregates	Water
	CEMENT PASTES AND MORTARS			
1	1784	0	0	446
2	1484	0	252	445
3	983	0	678	442
4	884	0	828	442
5	679	0	937	441
6	872	872	0	436
7	728	728	248	437
8	485	485	669	437
9	398	398	819	438
10	336	336	928	437
	CONCRETE			
Control	318	0	1910	191
Slag mix	159	159	1902	190

TABLE 4—*Regression equations of the form* $\sigma = A X^n$, *where* $\sigma(MPa)$ *is the cube strength and the volume concentration of hydrate is* X.

Material	Range of w_o	r	A	n	R^2
C_1/S_1	0.25 to 0.65	1.00	510	3.44	0.92
C_1/S_1	0.45	0.75	389	2.67	0.82
C_1/S_1	0.25 to 0.65	0.50	486	2.88	0.89
C_1/S_1	0.45	0.25	296	2.27	0.78
C_2	0.6	1.0	198	2.31	0.98
C_2/S_1	0.6	0.5	284	2.25	0.99
C_2/S_2	0.6	0.5	233	1.84	0.99
C_2/S_3	0.6	0.5	233	2.19	1.00
C_2/S_4	0.6	0.5	415	2.49	0.98
$C_2/$quartz	0.6	0.5	90	2.00	0.8

space not occupied by anhydrous cement. This approach is summarized in the Appendix.

Solid Specimens Hydrated in Pyknometer Bottles

Mixture proportions for the solid specimens are given in Table 3. The mixtures were cast in pyknometer bottles and were compacted by vibration under vacuum. At the same time, cubes — 50 mm for pastes and mortars, and 100 mm for concrete — were compacted under heavy vibration to within 1% of their calculated density. The cubes were crushed at ages of 3, 7, 14, 28, and 56 days. The results of these tests are plotted against the calculated volume concentration of product in Figs. 1, 2, and 3. The results of curve fitting of these data are given in Table 4.

Reference *13* affords a more conventional estimate of hydraulicity by comparing the strength of a mixed cement with one in which the BFS is replaced by the same mass of finely ground crystal-line quartzite which is, practically, inert. *H*, the hydraulicity index [*13*], for various BFSs were calculated for each of the concrete mixtures and these are related to corresponding volume concentrations of hydration product in Table 5. The relationship between these two parameters is approximately linear as shown in Fig. 4.

Differential Sorption of Water and Kerosene and Its Relationship to Drying Shrinkage and Creep

In the previous section, the quantity of the cement hydration products was specified by mass and absolute volume without regard to its dispersion in the available void space. Also the water evaporated at 110°C was treated as normal liquid water. In fact, a portion of this water is sufficiently strongly attracted to the solid [*14, 15*] that it must be treated as a temporary part of the solid [*16*].

In previous work [*17, 18*] it has been shown that the dried products of cement hydration, on re-saturation, may accommodate a greater volume of water than other liquids such as methanol. This effect has been ascribed to the differences

TABLE 5—*Hydration Index, H Ref* 13, *and hydrate concentration,* X, *for concrete mixtures made with various BFSs in* r = 0.5 *mixtures.*

	C_2/S_1		C_2/S_2		C_2/S_3		C_2/S_4	
Age Days	H	X	H	X	H	X	H	X
3	0.50	0.24	0.40	0.24	0.50	0.18	0.10	0.19
7	0.71	0.30	0.47	0.30	0.71	0.26	0.43	0.25
14	0.94	0.34	0.93	0.36	0.94	0.30	0.72	0.30
28	0.88	0.38	0.96	0.41	0.79	0.33	0.83	0.33
56	1.00	0.41	1.14	0.45	0.95	0.36	0.86	0.37

NOTE—Computation of hydraulicity factor (Ref *13*).

Suppose that it is required to assess the cementing quality of material Z when combined with PC in the mass proportions r units of PC and 1 − r units of Z. Suppose further that one reference mix is made with r mass units of PC and 1 − r mass units of crystalline quartz ground to cement fineness (I), and the other reference mix be made with 1.0 mass units of PC. It is assumed that the starting porosity is the same in all cases. The strengths at a particular age are as follows:

	Strength
For pure PC	σ_p
For mixture PC/Z	σ_z
For mixture PC/I	σ_i

The hydraulicity factor H is given by

$$H = \frac{\sigma_z - \sigma_i}{\sigma_p - \sigma_i}$$

If the material of interest is as good as PC the value of $H \geq 1$.

If the material of interest has no cementing value, it will yield the same strength as the inert reference material I, and, H = 0.

in the ratio of molar weight to density—a rough measure of molecular volume of the liquid [*18*]. Feldman, on the other hand, describes the phenomenon as rehydration of the cement paste [*19*].

If V_w is the volume fraction of the whole occupied by evaporable water, and V_L is that occupied by a liquid with large molecules, the difference $(V_w − V_L)$ represents the volume of "active" water which is called gel water or adsorbed water by Powers [*20*] and interlayer water by Feldman and Sereda [*21*]. In either case, one would expect the mass $(V_w − V_L)$ to be closely related to the volume of hydration product. It has also been linked to the volume of water held at low vapor pressure, and it is believed that the movement of this category of water is responsible for shrinkage and creep [*18, 22*].

For the present investigation, following experimental procedures given in Refs *18* and *22*, V_w was determined gravimetrically as the volume of water lost when an initially saturated specimen was oven-dried at 110°C. V_L was determined as the volume of kerosene sorbed when the oven-dry specimen was vacuum saturated at room temperature. $m = (V_w − V_L)/V_w$ thus represents the volume fraction of evaporable water which resides in space which is inaccessible to kerosene.

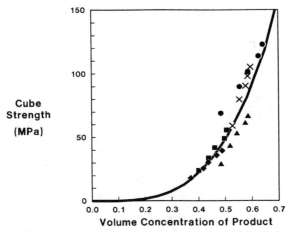

FIG. 1—*Dependence of cube strength, σ, on volume concentration, X, of hydration products in the space not occupied by anhydrous cement for PC pastes and mortars. Values of W/C ratio: $w_o = 0.25$ (●), 0.30 X, 0.45 (▲), 0.55 (■), and 0.65 (♦). Fitted curve is $\sigma = 510\ X^{3.44}$. Each set of five points corresponds to tests carried out at ages of 3, 7, 14, 28, and 56 days.*

Creep and shrinkage tests were carried out on concrete prisms measuring 100 by 100 by 700 mm. Creep specimens were loaded axially in pairs by means of prestressing steel rods in such a manner that each specimen containing a mixture of PC and BFS was stressed against a PC control specimen in the same rig. These specimens, together with unloaded shrinkage prisms, were exposed to the laboratory atmosphere for the test period. De-mountable strain gages were used to monitor total strain and shrinkage strain.

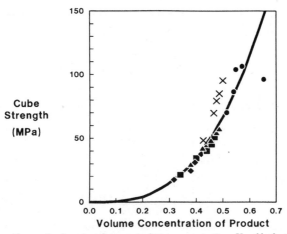

FIG. 2—*Dependence of cube strength, σ, on volume concentration, X, of hydration products in the space not occupied by anhydrous cement for PC/(PC + BFS) = 0.5. Pastes and mortars with $w_o = 0.25$ (●), 0.30 X, 0.45 (▲), 0.55 (■), and 0.65 (♦). Fitted curve is $\sigma = 486\ X^{2.88}$. Each set of five points corresponds to tests carried out at ages of 3, 7, 14, 28, and 56 days.*

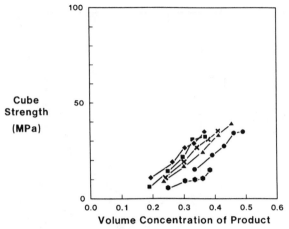

FIG. 3—*Cube strength of concrete made with aggregate : cement ratio = 6, W/C ratio = 0.6, and r = 0.50 combinations of PC C_2 and four different BFSs, S_1 X, S_2 (▲), S_3 (◆), and S_4 (■). Blank mix with milled quartzite (●). Each set of five points corresponds to tests carried out at ages of 3, 7, 14, 28, and 56 days.*

Table 6 gives the results of creep and shrinkage tests which lasted from 15 days age to approximately 1200 days. The same table gives the strengths at 14 days, m-values for kerosene and water, and $\Delta w/w_o$-values at 14 days. The correlation of m with strength is poor, but drying creep and shrinkage correlate well with 14 day values of m as shown in Fig. 5. It is clear that if m had been used to rank the PC—BFS mixtures with regard to potential volume change, the correct ranking would have been obtained.

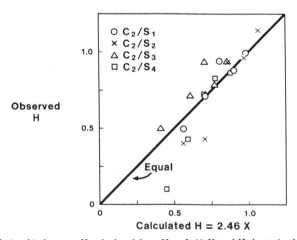

FIG. 4—*Relationship between H calculated from H = 2.46 X and H determined by experiment.*

TABLE 6 — *Values of* m *measured at age 14 days and values of creep and shrinkage measured between 15 and 1200 days.*

Mix $r =$	1.0	0.7	0.5	0.4	0.3
14 day strength, MPA					
$\sigma =$	28	31	29	30	25
$k* =$	0.104	0.140	0.174	0.198	0.241
$\Delta w/w_o =$	0.138	0.113	0.260	0.333	0.085
Shrinkage $\varepsilon_S =$ microstrain	440	500	570	620	630
Creep $\varepsilon_C =$ microstrain	1066	1540	1580	1910	2150

Discussion

Determination of Chemical Shrinkage

Measurement of chemical shrinkage requires great care, and, in particular, the hardened cement paste must have easy access to water. Autogenous shrinkage tests, such as those reported by Setter and Roy [3], are subject to error because the cement paste is isolated from the displacement medium by a rubber membrane. Under these circumstances, chemical shrinkage may cause water to retreat in a capillary which is bridged at the surface by the rubber membrane. A bubble then separates the capillary water surface from the membrane. The volume change sensed in the pyknometer bottle is in error by the sum of the volumes of the bubbles plus the volume expansion of the gross structure contained in the membrane.

If solid concrete is cast into a glass pyknometer bottle, expansion of the gross structure will burst the bottle in a few weeks. The test may be continued by

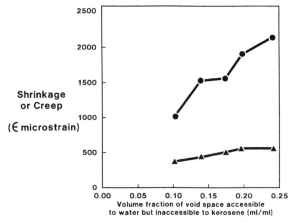

FIG.5 —*Drying creep (●) and shrinkage (▲), ε microstrain, ε related to volume fraction of voids* $k*$ *inaccessible to kerosene, where* $k* = (V_w - V_k)/V_w$, $V_w =$ *volume occupied by water and* $V_k =$ *volume occupied by kerosene. Each set of five points corresponds to tests carried out at ages of 3, 7, 14, 28, and 56 days.*

stripping the bottle and weighing the specimen which is then kept under water. The precision is much less than that obtained using the pyknometer bottle because the specimen has to be brought to a somewhat indeterminate saturated surface-dry condition each time it is weighed. It is possible to avoid damaging the bottle by separating the specimen from the walls by means of a porous bag or a corrugated metal cylinder. In either case it is necessary to apply vacuum each time the bottle is weighed to facilitate penetration of water into the solids.

As the specimens become increasingly dense, the rate at which water can permeate beyond the outer layers of material is less than the rate at which water is consumed by the chemical reactions. The specimens then self-dessicate even though they are continuously immersed in water [22]. Because of self-dessication the ratio of the mass of nonevaporable water to the mass of water taken up as hydration proceeds is not the same as it is in bottle hydrated samples. If the solid specimen were to be crushed and brought to a saturated surface-dry condition, the ratio should be the same as that obtained by bottle hydration.

In the present tests glass bottles and ceramic balls were used for ball-milling the cement slurries. The reaction between alkalis released by the cement and glass was not taken into account and is an obvious, though small, source of error which might have been avoided by the use of stainless steel pyknometer bottles.

Although comprising different species with different quantities of bound water, the volume of hydrate product is, on average, directly related to the mass of water which has reacted with the cement [2]. The mass of chemically bound water is related to the mass of water taken up to maintain saturation by the constant k. The quantity $(1 - 1/k)$ may be thought of as the specific volume of nonevaporable water, and it is a constant for a particular cement independent of the degree of hydration or the volume of hydration products.

The constant m represents the volume proportion of void space in the hydrate products which may be occupied by water and which excludes kerosene. It has been shown that this quantity of water is approximately that which is held in the paste at a relative humidity of 45% [18]; thus, according to Powers [26], it is proportional to the volume of hydrate. Fluids other than kerosene, some polar and some nonpolar, occupy different proportions of the total pore space [27], and the value of m, therefore, is specific to the fluid used.

The power curves fitted to the data may be compared with pooled results from Lawrence [23] and Roy [24] together with terminal values from Mills [25]. These are

	r	Equation	R^2
This investigation	1	$\sigma = 510\,X^{3.44}$	0.88
	0.5	$\sigma = 486\,X^{2.88}$	0.89
Pooled data from [23–25]	1	$\sigma = 536\,X^{2.30}$	0.99

NOTE—r = the ratio of PC to (PC + BFS).
  R = statistical coefficient of determination.

The implication of these equations is that the maximum attainable strength of hydrated cement is about 500 MPa whether PC or a mixture of PC and BFS is used. Roy [24] has measured strengths of 510 MPa for hot pressed PC pastes.

Summary and Conclusions

Although strength and W/C ratio are the most universally accepted indices of quality for cement paste, mortar, and concrete important qualities unrelated to these indices such as permeance and volume stability depend on porosity and X, the volume concentration of hydrate products.

Chemical shrinkage, as measured by the mass of water needed to make up the volume deficiency, is a useful parameter for estimation of X, the volume concentration of hydrate products. Unfortunately, the relationship between X and strength is not the same for all blends of portland cement and blast-furnace slag, although it follows the same form, the curves being parallel but displaced along the X-axis. A separate calibration is therefore necessary for each cement.

The differential sorption factor, m, was shown to be directly related to X for different cements and is, potentially, a useful parameter for estimating binder quality. For the same porosity, the higher the value of m, the higher the volume of water which is strongly bound to the solid and is therefore relatively immobile. The permeance of the material therefore decreases with increase in m.

On the other hand, both creep and shrinkage increase with m since the amount of "live" material in the cement paste is greater than similar material with lower m.

It appears that the hydration products formed from mixtures of portland cement and blast-furnace slag have a greater proportion of adsorbed or interlayer water than a pure portland cement and that this is responsible for the observed differences in both stressed and unstressed volume changes.

APPENDIX

Consider a hardened cement paste containing $(1 - \alpha)$ mass units of unhydrated cement. Assume a starting cement mass of unity and a starting W/C ratio w_o. If w_n^0 is the ratio of the mass of nonevaporable water in the completely hydrated cement to the mass of the cement, and Δw is the corresponding amount of water necessary to fill the space created by chemical shrinkage, we have

	Mass kg	Volume m^3/1000
Unhydrated cement	$(1 - \alpha)$	$(1 - \alpha)/d$
Hydrated cement	α	α/d
Nonevaporable water	αw_n^0	$\alpha w_n^0 - \Delta w$
Evaporable water	$w_o - \alpha w_n^0 + \Delta w$	$w_o - \alpha w_n^0 + \Delta w$
Sum	$1 + w_o + \Delta w$	$1/d + w_o$

Since the volume of hydration products is $[\alpha/d + \alpha w_n^0(1 - 1/k)]$, and the space available to accommodate it is $(w_o + \alpha/d)$, the volume concentration of product in the available space is given by

$$X = \alpha[k + dw_n^0(k - 1)]/k(dw_o + \alpha)$$

and, substituting $\alpha = w_n/w_n^0 = k\Delta w/w_n^0$, we get

$$X = \Delta w[k + dw_n^0(k - 1)]/[k\Delta w + dw_n^0 w_o]$$

For example:

For PC: $k = 3.87$; $w_n^0 = 0.253$; and $d = 3.22$ we get

$$X = 7.62\Delta w/(4.75\Delta w + w_o)$$

For $P/(P + S) = r = 0.5$; $k = 3.10$; $w_n^0 = 0.261$; $d = 3.09$ and

$$X = 4.79\Delta w/(3.1\Delta w + 0.8065w_o) = 5.94\Delta w/(3.84\Delta w + w_o).$$

References

[1] Copeland, L. E. in *Proceedings*, American Concrete Institute, Vol. 52, 1956, pp. 863–974; PCA Research Department Bulletin, No. 75, July 1956.

[2] Powers, T. C. in *The Chemistry of Cements*, H. F. W. Taylor, Ed., Academic Press, London, 1972, pp. 404–408.

[3] Setter, N. and Roy, D. M., *Cement and Concrete Research*, Vol. 8, 1978, pp. 623–634.

[4] Czernin, W., *Zementchemie fur Bauingenieure*, Bauverlag, Wiesbaden, 1960.

[5] Mills, R. H., *Transactions*, S. A. Institution of Civil Engineering, Vol. 4, No. 7, July 1962, pp. 125–132.

[6] Neville, A. M., *Properties of Concrete*, 3rd ed., Pitman, London, 1981, p. 373

[7] Powers, T. C., *The Chemistry of Cement*, H. F. W. Taylor, Ed., Academic Press, London, 1972, p. 414.

[8] Mills, R. H. in *Symposium on Structure of Portland Cement Paste and Concrete*, Highway Research Board, SR 90, Washington, 1966, pp. 406–424.

[9] Mills, R. H. in *Symposium on Structure of Portland Cement Paste and Concrete*, Highway Research Board, SR 90, Washington, 1966, p. 414.

[10] Czernin, W., *Zement und Beton*, Vol. 16, 1959, pp. 35–37.

[11] Copeland, L. E., *Proceedings*, American Concrete Institute, Vol. 52, 1956, pp. 863–874.

[12] Powers, T. C. in *The Chemistry of Cements*, H. F. W. Taylor, Ed., Academic Press, London, 1972, pp. 404–408.

[13] Lea and Desch, *The Chemistry of Cement*, 2nd ed., Edward Arnold, London, 1956, p. 407.

[14] Powers, T. C. in *Proceedings*, IV International Symposium Chemistry of Cement, Washington, 1960, p. 594.

[15] Derjaguin, B. V. and Abrikosova, I. I., *Journal of Physics and Chemistry Solids*, Vol. 5, 1958, pp. 1–10.

[16] Ramachandran, V. S., Feldman, R. F., and Beaudoin, J. J., *Concrete Science*, Heyden, London, 1981, pp. 56–72.

[17] Mikhail, R. S. and Selim, S. A. in *Symposium on Structure of Portland Cement Paste and Concrete*, Highway Research Board, SR 90, Washington, 1966, pp. 123–134.

[18] Mills, R. H. in *Proceedings*, V International Symposium on Chemistry of Cements, Part III, Tokyo 1968, pp. 74–85.

[19] Feldman, R. F. in *Special Publication SP-79*, American Concrete Association, Detroit, 1983, pp. 415–433.

[20] Powers, T. C. in *Proceedings*, IV International Symposium on the Chemistry of Cements, Washington, 1960, p. 602.

[21] Ramachandran, V. S., Feldman, R. F., and Beaudoin, J. J., *Concrete Science*, Heyden, London, 1981, pp. 62–70.

[22] Mills, R. H., *Transactions,* South American Institution of Civil Engineering, Vol. 11, Jan. 1969, pp. 1–11.

[23] Lawrence, C. D., "The Properties of Cement Paste Compacted Under High Pressure," Cement and Concrete Association Research Report 19, June 1969.

[24] Roy, D. M. and Gouda, G. R., *Journal of the American Ceramic Society,* Vol. 53, No. 10, 1973, pp. 549–50.

[25] Mills, R. H., *Symposium on Structures of Portland Cement Paste and Concrete,* Highway Research Board, SR 90, Washington, 1966, p. 410.

[26] Robertson, B. and Mills, R. H., *Cement and Concrete Research,* Vol. 15, 1985, pp. 225–232.

[27] Powers, T. C., and Brownyard, T. L., *Journal of the American Concrete Institute,* Vol. 18, No. 3, 1946, pp. 249–336.

Portland Fly Ash Cements

Nemat Tenoutasse [1] *and Anne-Marie Marion* [2]

Mechanism of Hydration of Cement Blended with Fly Ashes

REFERENCE: Tenoutasse, N. and Marion, A. M., **"Mechanism of Hydration of Cement Blended with Fly Ashes,"** *Blended Cements, ASTM STP 897,* G. Frohnsdorff, Ed., American Society for Testing and Materials, Philadelphia, 1986, pp. 65–85.

ABSTRACT: Selective dissolution of different Belgian fly ashes (FAs) with water, hydrochloric acid (HCl), and hydrofluoric acid (HF) solutions was investigated by chemical and microscopical techniques.

In regard to the use of fly ash for blended cement fabrication, the behavior of FAs was also studied in lime saturated solution.

The release of different oxides from FA in water, hydrochloric acid, hydrofluoric acid, and calcium hydroxide ($Ca(OH)_2$) solution, gives useful data about the influence of Belgian FA in blended cement hydration. The chemical analysis of liquid phase (from 8 h to 28 days) confirms the pozzolanic activity of FAs in lime saturated solution. Chemical data and microscopical examination demonstrate that all the sulfate and most of the potassium oxide (K_2O) content of the FAs particles are in the superficial layers, whereas sodium oxide (Na_2O) is dispersed in all the amorphous phase.

The influence of FA substitution on the microstructure of hydrated normal portland cement (OPC) pastes was also studied. FA act to increase the total porosity. After 4 months of hydration, the pore size distribution of systems with low FA content is rather similar to that of the pure paste.

The pozzolanic activity of FAs was investigated by microscopical study and free lime determination of the mixtures. A significant pozzolanic activity of FAs was observed after 14 days.

KEY WORDS: portland cement, fly ash, pore size distribution, mercury penetration

The use of fly ashes (FAs) in blended cement has been gaining increasing interest in view of energy saving. For about two years, a cement containing 20% of FAs has been produced in Belgium.

In our laboratory, we have begun a systematic study of the influence of FA substitution on the hydration of cement, especially in relation to the microstructure of cement blended pastes [1].

[1]REDCO, Research, Development, and Engineering Company, Brussels, Belgium.
[2]U.L.B., Department of Industrial and Solid State Chemistry, Brussels, Belgium.

TABLE 1 — *Chemical analysis and mineralogical composition.*

CHEMICAL ANALYSIS	
Ignition loss	0.57
Insoluble residue	0.11
SiO_2	19.99
Al_2O_3	5.44
Fe_2O_3	3.24
Total CaO	65.64
MgO	0.78
SO_3	2.07
Na_2O	0.30
K_2O	0.39
MINERALOGICAL COMPOSITION	
C_4AF	9.9
C_3A	8.9
C_2F	0.0
C_3S	60.1
C_2S	9.4
$CaCO_3$	0.6
$CaSO_4$	3.5

It is well known that the total porosity and the pore size distribution play a very important part in the durability of cement based materials. Indeed, the durability of concrete is greatly influenced by the permeability, which is, in turn, governed by the pore size distribution (2 to 4).

To achieve a meaningful characterization of FAs, the solubility behavior of their sulfates and alkali oxides has been measured in different conditions. These data are very useful for assessing the influence of the FAs on the "alkali-aggregate" reaction.

Materials and Methods

Table 1 gives the chemical analysis and mineralogical composition of cement (ASTM Type I) used in this work.

The specific surface of normal portland cement (OPC) was 2500 cm^2/g (ASTM Standard Test Method of Fineness of Portland Cement by Air Permeability Apparatus (C 204)). This is an industrial cement, produced in Belgium by the wet process. The setting regulator was natural gypsum. The chemical composition of typical Belgian FAs used in this investigation is presented in Table 2.

The hydration mechanism was investigated as a function of time, for OPC and OPC containing from 10 to 80% of FAs. In order to have a better understanding of the intrinsic behavior of the FAs, we also investigated the system "fly ash-lime," the major advantage of these experiments consists in the elimination of the interferences of minor oxides of cement (alkali oxides, sulfates, etc.)

For porosimetric and microscopic studies, 100 g of each specimen were prepared with distilled water corresponding to a 0.5 water/solid (W/S) ratio. After

TABLE 2—*Chemical analysis of FAs.*

	FA 1	FA 2	FA 3	FA 4
Loss	5.12	9.26	3.17	1.34
Insoluble	0.00	0.00	0.00	0.00
SiO_2 sol	52.20	46.95	53.55	52.10
Al_2O_3	26.44	24.62	27.55	28.92
Fe_2O_3	5.94	6.72	6.34	10.20
MnO	0.08	0.11	0.08	0.05
TiO_2	0.89	0.82	0.93	1.07
P_2O_5	0.26	0.53	0.28	0.34
Total CaO	2.41	3.91	1.40	1.59
MgO	1.51	1.98	1.57	1.17
SO_3	0.29	0.62	0.26	0.55
S	0.00	0.00	0.00	0.00
Cl	0.00	0.02	0.00	0.00
Na_2O	0.96	0.92	1.08	0.71
K_2O	3.28	2.56	3.20	2.21
Carbon	4.56	8.57	2.85	0.75
Free CaO	0.35	1.06	0.18	0.12

mixing by hand, the pastes were cast in cylindrical plastic vials, and capped to airtight conditions for 24 h before demolding. The demolded specimens were stored in airtight containers, until ready for testing.

After required hydration periods, 2 to 3 g pieces were cut from each specimen. Hydration was then stopped by alcohol-ether washing, and the remaining water was removed by drying the specimens at 70°C, under vacuum for 24 h.

Micromeritics equipment (pore size 9300) was used. The total pore volume and the pore size distribution were obtained by measuring mercury intrusion using pressures up to 206 MPa (30 000 psi). A mercury contact angle of 130° was selected. Different techniques, well known for the study of hydrated products were also used: X-rays diffraction, atomic absorption, and scanning electron microscopy.

Results and Discussion

Fly Ash Characterization

Chemical Analysis — All Belgian FAs are produced by the burning of bituminous coal. They could be classified in the Class F of ASTM Specification for Fly Ash and Raw or Calcined Natural Pozzolan for Use as a Mineral Admixture in Portland Cement Concrete (C 618-83). Indeed their sulfates and lime content is very low (as seen in Table 2). The solubility of FAs in pure water and acid solutions was investigated. The results of the water treatment are presented in Table 3.

The rate of sulfate release increases with the W/S ratio. Those chemical data, completed by microscopical examination allow us to locate all the sulfates in the external layers of the fly ash particles.

TABLE 3—*Solubility of FAs in H_2O (60 min, room temperature).*

		Dissolved Oxides, %			
		CaO	Na$_2$O	K$_2$O	SO$_3$
W/S = 2	FA 1	5.0	2.0	1.1	100
	FA 2	11.0	2.5	0.9	35.7
	FA 3	8.0	1.2	0.5	100
	FA 4	12.2	1.8	0.6	79.5
W/S = 10	FA 1	10.6	1.9	1.3	100
	FA 2	12.0	2.2	0.9	100
	FA 3	12.0	1.0	0.6	94
	FA 4	18.0	1.9	0.7	100

The comparison of data obtained for the solubility of calcium and sulfates in aqueous phase confirm the presence of calcium sulfate (anhydrite) in FAs. We also observe, even with a relatively high W/S ratio, the very low solubility of FA alkali oxides in distilled water. In hydrochloric solution (Table 4), FAs release all of their sulfates and a significant quantity of calcium oxide (CaO). The dissolution of alkali oxides remains weak.

FAs are seriously attacked by hydrofluoric acid (HF), as shown in Fig. 1. We observe a quick release of potassium oxide (K_2O) in HF: after 15 min, all the K_2O content is released. During the same period, the FA releases only 45% of its sodium oxide (Na_2O) content. These data suggest a different location for the two alkali oxides.

Microscopical Study —The FA residues after water and acid treatments were examined by SEM. Concerning the water and hydrochloric acid (HCl) treatment, there does not appear any deep change in the morphological aspect of the particles (Fig. 2). The conclusion of this microscopical observation, supported by chemical data is rather evident; all the sulfates are dissolved without any structural damage to the particles. This fact confirms the location of the totality of the sulfates at the surface of the FA particles.

The drastic HF attack of the FA particles is demonstrated by the scanning electron microscope (SEM): amorphous silica phases (glasses containing relatively high contents of K_2O) are dissolved, and it is easy to observe the cristalline phases of FAs (Figs. 3a and 3b), just below the glass surface. The SEM and

TABLE 4—*Attack of FAs by 0.1 N HCl, 60 min, room temperature.*

		Dissolved Oxides, %				
		CaO	Na$_2$O	K$_2$O	SiO$_2$	SO$_3$
W/S = 25	FA 1	68.0	5.9	2.9	0.3	100
	FA 2	86.0	3.4	2.3	0.2	99.2
	FA 3	57.7	1.5	1.1	0.1	100
	FA 4	55.8	4.3	2.0	0.2	95.0

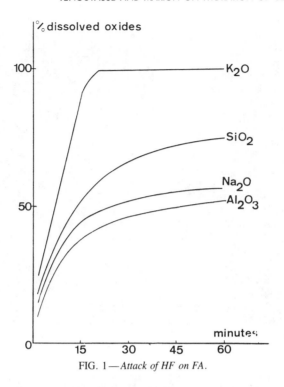

FIG. 1—*Attack of HF on FA.*

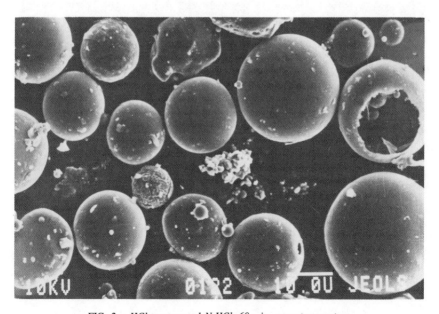

FIG. 2—*HCl treatment, 1 N HCl, 60 min, room temperature.*

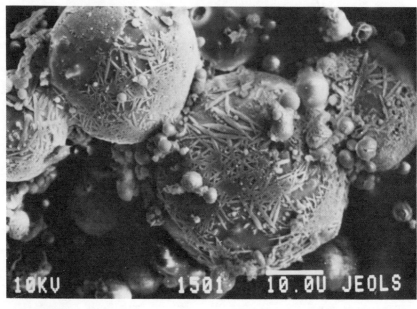

(a)

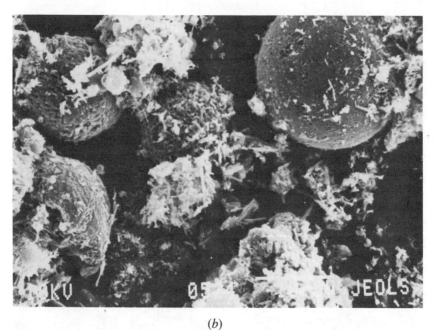

(b)

FIG. 3—*HF treatment, 0.5 N HF, (a) 2 min, room temperature and (b) 3 h, room temperature.*

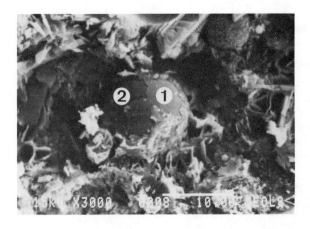

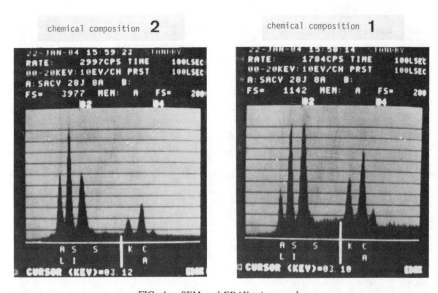

FIG. 4—*SEM and EDAX micrographs.*

energy dispersive analyses by X-ray (EDAX) micrographes, presented in Fig. 4, confirm the location of potassium in the external layer of the FA particles.

Pozzolanic Properties of Fly Ashes

FA-Lime System —The study of the FA-lime system shows the low solubility of the alkali oxides from FAs, compared with the cement. In this above system, the evolution of calcium concentration is quite significant in regard to the pozzolanic activity of FAs. As shown in Fig. 5, the calcium content in the liquid

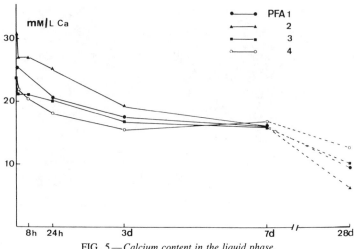

FIG. 5 — *Calcium content in the liquid phase.*

phase decreases significantly between 8 h and 28 days. Those chemical data suggest a pozzolanic reaction between the FA particle and the calcium hydroxide $Ca(OH)_2$; indeed the system contains an excess of solid $Ca(OH)_2$, sufficient to saturate the liquid phase, if there is no calcium hydrosilicates (CSH) formation.

After 28 days treatment with $Ca(OH)_2$, the morphological aspect of some FA particles is rather similar to that observed after HF treatment (Fig. 6).

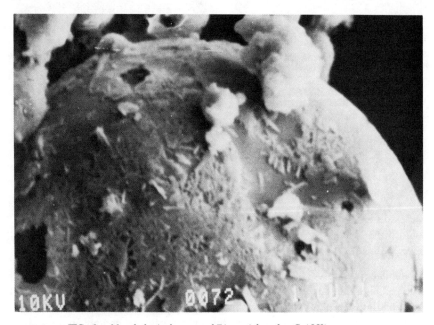

FIG. 6 — *Morphological aspect of FA particles after Ca(OH)$_2$ treatment.*

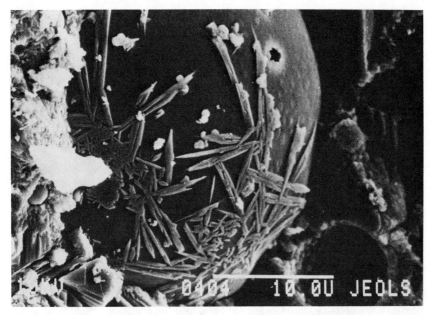

FIG. 7—*Present FA particles after 60 days hydration.*

OPC-FA System — The pozzolanic properties of FAs are also demonstrated by the determination of the free lime content of the samples hydrated in the systems OPC and OPC containing from 10 to 80% FA. This effect is more significant for the systems with 60 and 80% PFA where the pozzolanic activity of FAs is clearly demonstrated after only 14 days of hydration.

The SEM study of the system OPC plus FAs exhibits the morphological aspect of the pozzolanic activity developed by the FA particles during hydration of cement blended pastes. Figures 7 and 8 present FA particles after 60 days hydration, with typical mullite crystals.

Figure 9 is interesting in regard to the heterogeneity of the attack of the FA particles, (one of the particles seems to be intact, even after 60 days of hydration), which seems to depend upon the intrinsic reactivity of each particle. Figures 10 and 11 illustrate the pozzolanic reaction of FAs after 120 days of hydration: some macropores become apparent at the surface of the particles. Those macropores are of course formed during the cooling process of FAs, and generally they are not observable at the surface of unattacked particles (Fig. 12). They appear after erosion of external layers of the FA particles (Fig. 13).

After 150 days, some particles are seriously attacked; after dissolution of the amorphous silica phase, they break down into their own constitutive mullite crystals, as it can be seen in Fig. 14.

Porosimetric Data — The porosimetric data are expressed as $(\triangle V \times 100)/(\triangle V_{max})$ as a function of the mean pore diameter. Figure 15 presents the pore size

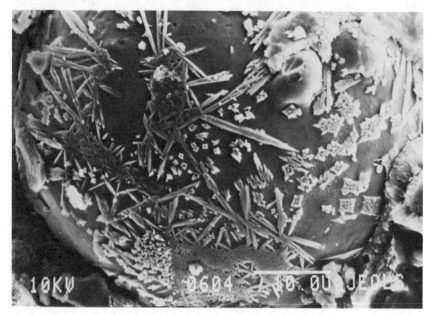

FIG. 8—*Present FA particles after 60 days hydration.*

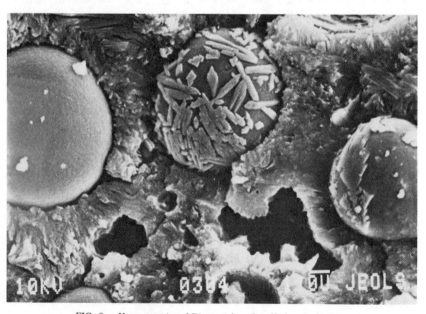

FIG. 9—*Heterogeneity of FA particles after 60 days hydration.*

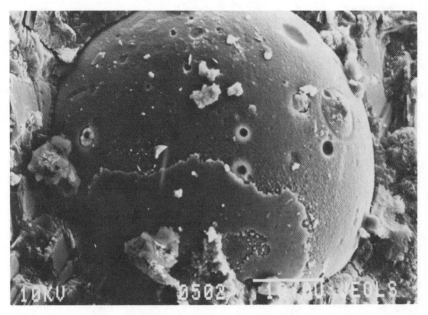

FIG. 10—*Pozzolanic reaction of FA after 120 days of hydration.*

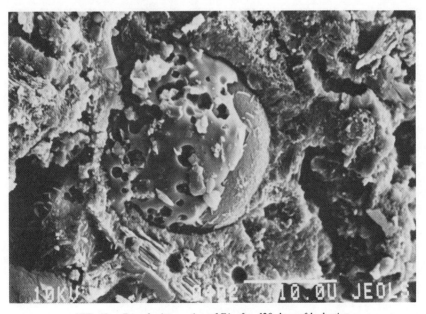

FIG. 11—*Pozzolanic reaction of FA after 120 days of hydration.*

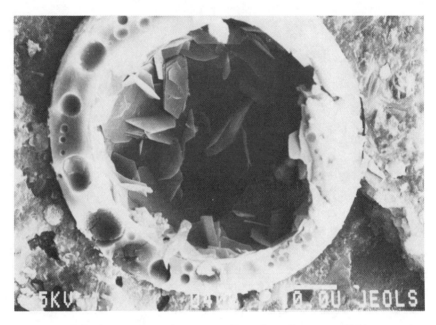

FIG. 12—*Formation of macropores during the cooling process of FA.*

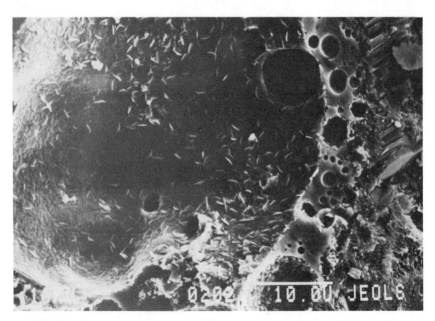

FIG. 13—*Macropores as they appear after erosion of external layers of FA particles.*

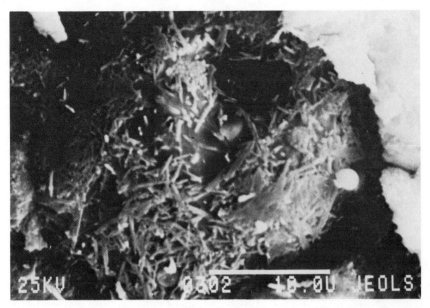

FIG. 14—*Breakdown into constitutive mullite crystals after 150 days.*

distribution of pure OPC paste hydrated from 3 days to 1 year. The total porosity decreases till 9 months, and the pore size distribution is shifted to smaller diameter versus of hydration time.

A substitution of 10 to 20% of OPC by FAs does not change the development of the microstructure, but produces a significant increase of the total porosity, especially for the early hydration periods (Figs. 16 and 17).

After 5 months hydration, the pore size distributions obtained for OPC and OPC blended with 10 to 20% FA 4 are very similar. Porosimetric data of OPC containing 10 to 20% FA 2 support these results.

The pore size distribution of systems with high FA content (80%) are quite different (Figs. 18 and 19).

A significant porosity is observed between 0.1 and 10 μm, probably due to unreacted FA particles. Indeed, voids exist between FA particles; in the system with high FA content, this interparticular pore volume is important, and it can not be entirely filled with hydration products, because of the poor cement content (20%). It is then easily understandable that mercury fills those pores during intrusion. The various amount of pore observed in this region (between 0.1 and 10 μm) for the two FAs, suggests an eventual difference in the pozzolanic reactivity.

For the early hydration period, the microstructure of these pastes remains poorly developed due to their low cement content. However, after 28 days' hydration, the micropores grow significantly, with a narrow pore size distribution, when compared with pure OPC pastes. The hydrosilicates formed by the

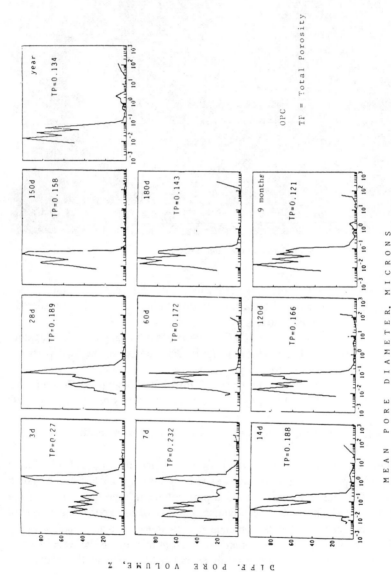

MEAN PORE DIAMETER, MICRONS

FIG. 15—*Pore size distribution of pure OPC paste hydrated from 3 days to 1 year.*

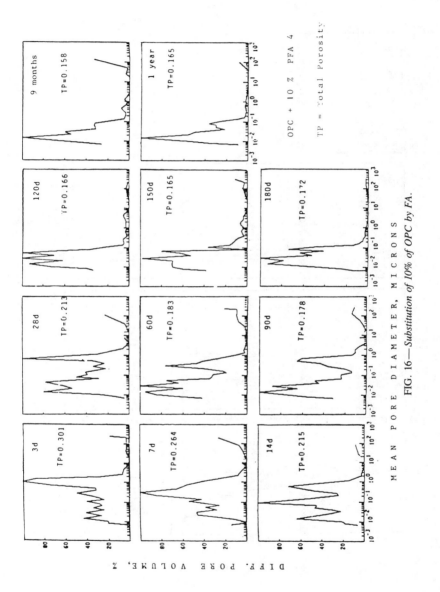

FIG. 16—*Substitution of 10% of OPC by FA.*

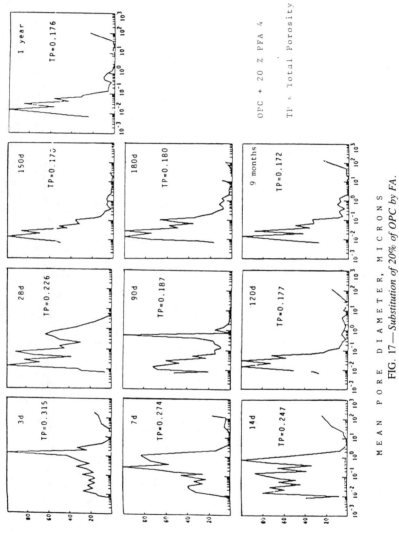

MEAN PORE DIAMETER, MICRONS

FIG. 17 — *Substitution of 20% of OPC by FA.*

DIFF. PORE VOLUME, %

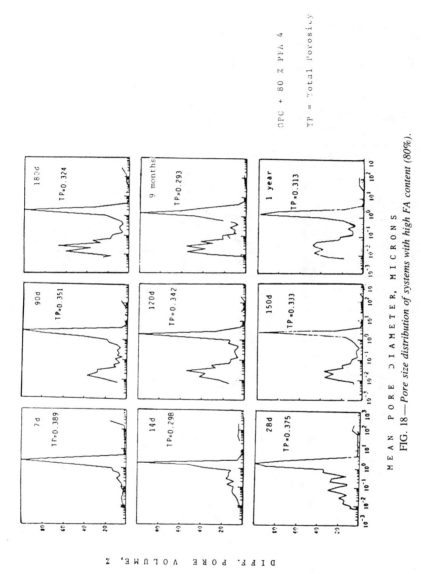

FIG. 18—Pore size distribution of systems with high FA content (80%).

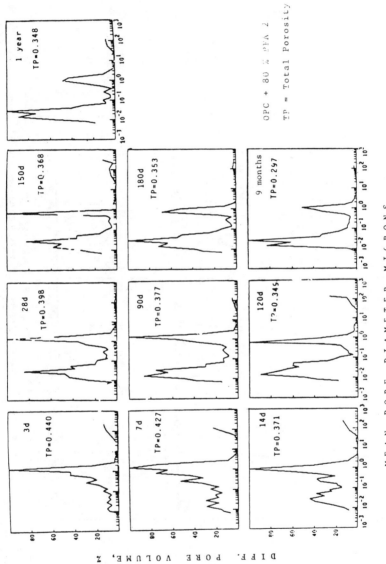

MEAN PORE DIAMETER, MICRONS

DIFF. PORE VOLUME, %

FIG. 19—*Pore size distribution of systems with high FA content (80%).*

TABLE 5 — Free lime content (W/S = 0.5).

Specimen	1 day	2 days	3 days	7 days	14 days	28 days	60 days	90 days	120 days	150 days	180 days	1 year
OPC 100%	7.13	9.12	10.69	12.89	15.60	13.89	16.025	16.11	16.32	16.09	15.17	13.55
OPC + 10% FA 2	6.81	8.22	10.95	12.40	13.22	14.70	13.77	14.22	15.67	14.46	11.66	9.32
+ 20% FA 2	5.95	8.45	10.17	11	12.99	13.02	12.28	12.79	13.54	12.77	11.34	6.74
+ 30% FA 2	5.59	8.15	8.77	9.95	11.56	11.90	11.01	10.66	11.15	10.6	8.46	4.31
+ 40% FA 2	5.14	7.37	7.58	9.65	9.90	10.88	9.85	9.48	9.08	8.62	8.13	4.08
+ 60% FA 2	3.85	5.14	5.32	7.14	7.06	6.52	5.87	5.31	4.58	4.92	4.06	1.82
+ 80% FA 2	3.08	2.43	3.59	3.65	3.39	2.72	2.08	2.41	2.35	1.86	1.56	0.93
OPC + 10% FA 4	6.01	8.31	8.39	11.42	13.43	13.93	14.40	13.76	15.19	13.26	12.65	9.80
+ 20% FA 4	5.90	7.67	9.56	11.55	13.26	11.43	11.99	12.72	12.96	12.61	11.97	6.39
+ 30% FA 4	5.13	7.14	8.27	9.29	10.37	11.44	10.96	10.23	10.95	10.08	10.08	4.89
+ 40% FA 4	4.29	6.24	7.41	9.37	9.67	10.95	8.52	9.05	8.42	8.38	7.56	4.28
+ 60% FA 4	2.92	4.19	5.02	6.55	7.01	6.21	5.77	5.6	4.45	3.85	3.80	1.59
+ 80% FA 4	1.03	1.89	2.49	3.38	3.41	2.78	1.78	1.75	1.39	1.32	1.28	0.65

pozzolanic reaction between FAs and $Ca(OH)_2$ generated by cement hydration are probably responsible for this microstructure. Indeed, the pozzolanic reactivity of FAs has been clearly demonstrated, from 28 days of hydration (SEM, free lime content, calcium concentration in aqueous phase).

Summary and Conclusions

Investigations on the solubility of FA in different conditions demonstrate the presence of the totality of the sulfates in the external layers of FA particles. The data from acid treatment, supported by SEM observations allow us to conclude to a different location of Na_2O and K_2O.

The pozzolanic activity of FA is confirmed by the study of the behavior of FA in lime saturated solution. The determination of free lime in OPC and OPC plus FA pastes suggests a reaction between $Ca(OH)_2$ and FA particles (after 1 year, a significant amount of $Ca(OH)_2$ is combined with FAs).

The study of pore size distribution of cement blended with FAs shows:

1. The addition of FAs to OPC always increases the total porosity, especially for the early hydration period.

2. For cement blended with reasonable amount of FA (up to 25%), the total porosity and pore size distribution after 3 months hydration are rather similar to those of OPC paste.

These results are very useful to surface characteristic determination of FAs.

As it is well known [5], the surface properties of FAs play an important part in engineering properties of concrete (water demand, workability, etc.). Indeed, according to some authors, FAs generally increase the water demand of cement [6].

Acknowledgment

This paper reports results that are a part of a research contract between CRIC[3] and IRSIA[4] devoted to the study of blended cements.

The authors gratefully acknowledge IRSIA for their financial support. Thanks are also due to D. Van Wouwe of the central laboratory of Redco, for developing a software for processing of Hg-porosimetric data.

References

[1] Tenoutasse, N. and Marion, A. M., in *Proceedings,* British Ceramic Society meeting, Basic Science Section, Chemistry and Chemically Related Properties of Cement, London, April 1984, pp. 359–374.
[2] Mehta, P. K. and Manmohan, D., in *Proceedings,* 7th International Congress on Chemistry and Cements, Vol. 3, Paris, 1980, pp. VII–15.

[3]Centre National de Recherches Scientifiques et Techniques pour l'Industrie Cimentière, Bruxelles, Belgium.
[4]Institut pour l'Encouragement de la Recherche Scientifique dans l'Industrie et l'Agriculture, Bruxelles, Belgium.

[3] Manmohan, D. and Mehta, P. K., *Cement, Concrete, and Aggregates,* Vol. 3, No. 1, Summer 1981, pp. 63–67.
[4] Feldman, R. F., *Journal of American Ceramic Society,* Vol. 67, No. 1, Jan. 1984.
[5] Mehta, P. K., in *Proceedings,* First International Conference on the Use of Fly Ash, Silica Fume, Slag, and Other Mineral By-Products in Concrete, Canada, 31 July–5 Aug., 1983, Vol. I. pp. 1–46.
[6] Costa, V. and Massaza, F., in *Proceedings,* First International Conference on the Use of Fly Ash, Silica Fume, Slag, and Other Mineral By-Products in Concrete, Canada, 31 July–5 Aug., 1983, Vol. I., pp. 235–254.

Blended Cements with Slag and Pozzolans

Ronald H. Mills [1]

Evaluation of the Performance of Blast-Furnace Slag and Fly Ash When Blended or Mixed with Portland Cement

REFERENCE: Mills, R. H., "Evaluation of the Performance of Blast-Furnace Slag and Fly Ash When Blended or Mixed with Portland Cement," *Blended Cement, ASTM STP 897,* G. Frohnsdorff, Ed., American Society for Testing and Materials, Philadelphia, 1986, pp. 89–105.

ABSTRACT: Partial substitution of portland cement (PC) with materials such as blast-furnace slag (BFS) or fly ash (FA) affects: workability or water demand; characteristic strength at a given age; and the maturity needed to attain a given strength.

Assuming constant workability, the efficiency of the substitute material may be stated in terms of the ratio of the mass of BFS plus FA to the mass of PC which is substituted.

For equal workability and strength, the efficiency factor varies with blend proportions and the characteristic strength. In the present investigation, the efficiency factor diminished for all strength levels as the proportion of BFS increased from 25 to 75%. The maturity efficiency factor is given by the ratio of log (maturity) values for the cements being considered. In this case the mass ratio water/binder is constant. Maturity efficiency factors reduced with increase in BFS but did not vary with age.

KEY WORDS: blast-furnace slag, cement, fly ash, portland cement, hydration, strength, efficiency, maturity

The most widely quoted incentives for the use of blast-furnace slag (BFS) and fly ash (FA) in part substitution for portland cement (PC) are concerned with economy of the final concrete mix, either in money or in energy. Other reasons for using PFS or FA which may not be so obvious include the following:

1. Improved resistance to sulfate in soil or groundwater [1]. This is the case for PC/BFS blends but not necessarily correct for PC/FA blends.

2. Reduction of seasonal variability of cement from individual factories and between different factories by appropriate additions of BFS and FA to clinker [2]. Such composite cements are designed to have more carefully controlled characteristics than PC.

---

[1]Professor of civil engineering, University of Toronto, Toronto, Ont. Canada.

3. Improved security against alkali-aggregate reaction [3]. Since both BFS and FA are glassy materials containing silica in a disordered state, they offer competitive demands for calcium hydroxide, sodium, and potassium ions released into solution by hydration of PC. In some areas where Type 50 PC is not available, a PC/BFS blend may be mandatory.

Some factors which mitigate against the use of BFS or FA as cement components in concrete are:

1. Increased susceptibility to carbonation [4]. Thus blended cements should be avoided in concrete exposed to heavy industrial atmospheres.

2. Slower hardening, which requires prolonged moist curing.

3. In the case of cement substitutes added at the mixer, the need to procure, store, and batch an extra ingredient leads to increased risk of error or increased costs of supervision [5].

4. Over grinding of PC when milled together with harder BFS [6].

Characterization of the PC Substitute in Blends or Mixtures of PC/BFS or PC/FA by Performance in Concrete

Strength

Since many of the desired characteristics of concrete go hand in hand with strength, this parameter forms the basis for consumer comparisons between PC and PC/BFS or PC/FA. The objective is to design a blend or mixture which will produce concrete of the same strength as the parent PC. This is generally impossible to achieve unless appropriate boundary conditions are established. These boundary conditions may be stated as follows:

1. Strength, means the target strength σ which is necessary to ensure a given statistical probability that the specified strength f_c' will be met [7].

2. Furthermore, the curing regime should be specified as: either maturity at a standard temperature or, in special cases, at the service temperature.

3. Since σ rather than f_c' is adopted as the criterion, the standard deviation of concrete made with alternative cements should be considered.

4. The degree of moist curing should be considered because blended or mixed cements are generally slower to harden than PC.

In this paper, equal target strength at 28 days age was adopted as the strength criterion. Strength is generally governed by the quality of the cement paste, that is, the volume ratio of cement in the cement paste at the time of mixing. This is often expressed in terms of the mass ratio of water to cement and sometimes by the volume of water plus air to the volume of cement.

Workability

Workability depends primarily on the volume ratio of cement paste in the concrete and also the viscosity of the cement paste. The user is concerned with the investment of mechanical work required to achieve full compaction. This is usually judged by means of the slump test. In this paper workability was normalized to a slump range of 50 to 100 mm, corresponding to medium workability.

The characteristic water content necessary to achieve given workability usually is called water demand and expressed in kilograms per cubic metre of finished concrete. For purposes of comparing concretes with different air contents the initial void ratio = volume of (water + air) per cubic metre is used and denoted by V.

Economic Interdependence of Workability and Strength

Experimental data may be fitted to a logarithmic curve:

$$w_o = \alpha + \beta \log (f_c')$$

where

w_o = volume ratio of (water + air) to cement,
f_c' = cylinder strength, and
α, β = constants to be determined by experiment.

For a given value of f_c', w_o may be estimated from the empirical relationship. Furthermore, workability is a function of the volume ratio of paste V_p and its viscosity related to w_o.

Experimental data may be fitted to a power curve

$$V_p = A(w_o)^B$$

where

V_p = volume ratio of cement paste,
w_o = ratio of (water + air) to cement; and
A, B = empirical constants.

For a given w_o, V_p and hence the masses of cementitious materials may be estimated. An example of the treatment of experimental data is given in Appendix I. It is noted that the cement content = (water + air)/w_o. Thus, the cost of binder in concrete increases with both workability and strength.

Influence of Variability

To ensure an acceptable small risk of failure to meet specifications, both V_p and $1/w_o$ have to be set to average values higher than those required to meet the specification. Thus as the standard deviation in measured strength or workability increases, cement content, and hence cost, increases.

Efficiency Factors

In order to compare the cementing qualities of a material used in part substitution for PC in a blend or mix, the hydraulicity factor [8] may be used as outlined in Appendix II. The method compares the strength of concrete or mortar made with the blended cement, with a mixture of pure PC and with a mixture of PC and inert rock flour. Comparable mass proportions are maintained. The disadvantage of this method is that any difference in workability arising from characteristics of the cement substitute is ignored.

Warriss [9] has proposed an efficiency factor based on the rate at which calcium hydroxide released by the PC component is bound by the substitute material. This approach also does not take workability into account.

The practitioner is interested in a comparison of alternative concretes having equal workability and 28 day strength, one made with pure PC (P kg m^{-3}) and the other made with a blend or mix cement (C kg m^{-3}) in which the proportion of PC is r and of a cement substitute X is $1 - r$ [9, 10].

Common to both cements is a mass of PC = rC. The mass of material $X = (1 - r)C$ has to provide the same cementing value as $P - rC$.

The efficiency factor on a mass-strength basis is the mass ratio

$$\eta = (P - rC)/(1 - r)C$$

Although strength based on the lunar month is entrenched firmly as a measure of concrete quality, the practitioner is also interested in the time taken to attain a given strength, not necessarily the 28 day strength, at different temperatures.

The treatment outlined in Appendix III uses, as a quality parameter, the maturity to reach a given strength. The maturity efficiency of a Mix A containing a blended cement compared with Mix B containing only PC is defined as

$$\eta_m = (\log M_B / (\log M_A)$$

where M_A and M_B are, respectively, the number of degree hours needed for each mix to attain the specified strength level. On the Celsius scale M = the area under the time-temperature curve + 10(time). Time is usually measured in hours from the time of final set.

A fundamental difference exists between the two efficiency factors just mentioned. This may be summarized as follows:

Efficiency Factor	Maturity	Strength	w_o
η	fixed	fixed	variable
η_m	variable	fixed	fixed

A useful property of η_m is that it may be used to estimate the time-temperature regime needed to achieve a given strength or to achieve parity with a PC control.

Experimental

Data from Mills [11, 12], Malhotra [13], and Munday et al [14] were used to generate efficiency factors.

Materials

Oxide compositions and physical characteristics of PCs, BFSs, and FA are given in Table 1.

TABLE 1 — *Oxide composition and fineness of cementitious materials.*

Ref	Material	CaO	SiO$_2$	Al$_2$O$_3$	MgO	Fe$_2$O$_3$	SO$_2$	Blaine Fineness, m^2 kg^{-1}	Density
11	PC	63.8	21.3	4.5	2.6	2.0	3.2	340	3.22
	BFS	39.2	33.8	6.2	16.5	<1.0	1.9	420	2.90
12	PC	63.1	20.2	5.9	2.6	1.9	3.3	320	3.22
	BFS	38.0	36.9	7.5	12.3	<1.0	3.2	360	2.90
	FA	2.8	42.3	23.0	0.96	14.7	0.87	381	2.31
13	PC	62.7	21.8	4.5	2.5	2.2	3.2	373	...
	BFS	37.8	37.3	8.2	10.9	1.2	2.0	465	...

Concrete Mixes

Mixes of Ref *11* were adjusted to a predetermined workability by varying the ratio of cement paste to aggregate within a narrow range about a mean value of 30% by volume for each water:cement (W/C) ratio.

Mixes of Ref *12* were adjusted to a fixed water:solid (W/S) ratio of 0.23 in the fresh mix. Workability was adjusted to stiff plastic consistency (*VB°*) by additions of superplasticizer powder. These mixes were subjected to heavy compaction to eliminate air-filled voids. Mixes of Ref *13* were batched to give the same ratio of $W/(P + B)$ in each batch with approximately the same aggregate:cement ratio. Workability was maintained at the medium level for mixes containing an air entraining agent (AEA) but no superplasticizer (SP), and at a high level for mixes containing both AEA and SP.

Results

Mixes Containing Portland Cement and Blast-Furnace Slag

Data from Malhotra [*13*] and Mills [*11*] was processed by the methods outlined in Appendix I. Table 2 gives the constants of log curves fitted to w_o and f'_c. The corresponding values of w_o for values of f'_c from 20 to 40 MPa are given in Table 3. Although different materials and mixes were used in Refs *11* and *13*, there is good agreement in the 20 to 30 MPa range.

The coefficients of determination, R^2, generally supported the validity of the logarithmic curve fit.

Table 4 gives the coefficients in a power curve fit of data connecting the volume of paste V_p with w_o. Except for the 55/45 mixes with superplasticizer from Ref *13* the coefficients of determination reflect satisfactory goodness of fit.

Efficiency values determined from these data are shown in Table 5. As a crude approximation the efficiency value η may be regarded as a "break-even" cost factor. The cost ratio of BFS, as delivered to the consumers site, must be less than η for economic viability. The cost ratio of BFS to PC is more than 0.8 in Ontario and 0.6 in South Africa [*15*].

TABLE 2—*Values of constants in curves fitted to experimental data,* $w_o = \alpha + \beta \log (f'_c)$.

Ref	Age days	r	α	β	R^2
11	7	1.00	2.16	−1.04	1.00
		0.75	1.95	−0.91	0.99
		0.50	1.81	−0.90	0.98
		0.25	1.48	−0.80	0.99
	28	1.00	2.40	−1.10	1.00
		0.75	2.25	−1.09	0.98
		0.50	2.25	−1.09	0.98
		0.25	2.09	−1.05	0.98
13	7	1.00	2.51	−1.28	0.94
		0.75	2.53	−1.32	0.78
		0.55	2.30	−1.20	0.84
		0.35	1.94	−1.02	0.79
	28	1.00	2.77	−1.37	0.97
		0.75	2.90	−1.48	0.97
		0.55	2.76	−1.41	0.96
		0.35	3.09	−1.68	0.86

NOTE—r = mass ratio PC/(PC + BFS).
r = coefficient of determination.

It is seen that quite different conclusions as to the cementing value of BFS would be reached from the two studies [11, 13] even though the source of slag was the same. This is to be expected because the performance characteristics of the two groups of mixes were different. Solid superplasticizer was used in mixes of Ref 11 as a water reducer rather than to improve workability. These mixes also did not contain entrained air and the cylinders were fully compacted by vibration.

Table 6 gives the results of those mixes from Ref 13 which were dosed with superplasticizer in order to obtain high slump concrete. The values of η are very low and suggest that superplasticizer should not be used to obtain high workability in mixes containing BFS.

Influence of Variability

Another aspect of Table 5 which requires further study is the difference in η factors between the f'_c group and the $\sigma = (f'_c + 1.4 \ s)$ groups. The Ref 11 mixes have higher η factors when variability is taken into account because the s-values for PC were higher than for PC + BFS. For the data of Ref 13 an average s value was used for all comparable mixes. In the latter case η-values decreased between those determined from f'_c and those determined from σ.

In order to establish a range for values of η for a practical range of controls, the approach given in Appendix II was pursued. The results summarized in Table 7 are paradoxical in that η increases for $r = 0.75$, remains static for $r = 0.50$, and reduces for $r = 0.25$ as the degree of control degenerates from very good to poor.

TABLE 3—*Comparative values of w_o calculated from regression equations of Table 2 fitted to data from two references.*

28 Days	Ref 11, w_o for various r				Ref 13, w_o for various r			
f'_c	$r = 1.0$	$r = 0.75$	$r = 0.5$	$r = 0.25$	$r = 1.0$	$r = 0.75$	$r = 0.55$	$r = 0.35$
20	0.97	0.95	0.88	0.72	0.99	0.98	0.93	0.91
25	0.86	0.85	0.78	0.62	0.86	0.84	0.79	0.74
30	0.78	0.77	0.70	0.54	0.75	0.72	0.68	0.61
35	0.70	0.71	0.63	0.47	0.66	0.62	0.58	0.50
40	0.64	0.65	0.57	0.41	0.58	0.54	0.50	0.40
45	0.58	0.60	0.51	0.35	0.51	0.46	0.43	0.31

TABLE 4—*Values of* A *and* B *in curves fitted to experimental data,* $V_p = A(w_o)^B$.

Ref	r	A	B	R^2
10, with SP water reducer	1.00	232	−0.239	0.99
	0.75	229	−0.248	0.99
	0.50	232	−0.196	0.90
	0.25	208	−0.343	0.99
13, without SP	1.00	268	−0.366	0.99
	0.75	272	−0.329	1.00
	0.55	263	−0.435	0.99
	0.35	279	−0.400	0.87
13, with SP	1.00	260	−0.510	0.97
	0.75	255	−0.529	1.00
	0.55	254	−0.610	0.57
	0.35	231	−0.774	1.00

NOTE—r = mass ratio PC/(PC + BFS).
R = coefficient of determination.

TABLE 5—*Efficiency values for mixes having various strengths* f_c'; *target strengths* $\sigma = f_c' + 1.4(s)$ *and comparable workability.*

	Ref *11* for Medium Workability			Ref *13* for Low to Medium Workability		
r	0.75	0.50	0.25	0.75	0.55	0.35
$f_c' = 20$	0.98	0.85	0.78	0.92	0.91	0.83
25	1.02	0.86	0.73	0.89	0.84	0.73
30	1.02	0.87	0.70	0.85	0.80	0.66
35	0.97	0.87	0.66	0.78	0.72	0.58
40	1.09	0.88	0.62	0.77	0.66	0.49
45	1.17	0.87	0.57	0.71	0.62	0.39
$\sigma = 20$	1.02	0.88	0.78	0.87	0.82	0.79
25	1.09	0.98	0.71	0.83	0.74	0.71
30	1.15	0.89	0.70	0.78	0.68	0.61
35	1.17	0.88	0.64	0.72	0.63	0.52
40	1.20	0.90	0.61	0.63	0.53	0.42
45	1.22	0.93	0.57	0.60	0.47	0.34

NOTE—Efficiency $\eta = (P - rC)/(1 - r)(C)$; where P = PC content when $r = 1$; and C is the sum of PC and BFS in the blend.

TABLE 6—*Efficiency values for various* f_c' *and* σ *for high workability mixes with both air entrainment and superplasticizer* [13].

r	0.75	0.55	0.35
$f_c' = 30$	0.89	0.75	0.68
35	0.77	0.66	0.54
40	0.74	0.59	0.39
$\sigma = 30$	0.81	0.62	0.60
35	0.70	0.56	0.46
40	0.57	0.45	0.30

TABLE 7 — *Efficiency factor for* $f_c' = 30$ *MPa and three degrees of control.*

		Efficiency Factor η for Various r		
Control	$v\%$	$r = 0.75$	$r = 0.50$	$r = 0.25$
Very good	11	1.06	0.86	0.97
Fair	17	1.12	0.87	0.63
Poor	26	1.11	0.86	0.57

TABLE 8 — *Efficiency factors for BFS at greater age (Ref 13).*

	Efficiency Factor η					
	Age = 91 days			Age = 365 days		
f_c' (MPa)	$r = 0.75$	$r = 0.55$	$r = 0.35$	$r = 0.75$	$r = 0.55$	$r = 0.35$
	WITHOUT SUPERPLASTICIZER					
35	0.91	0.89	0.70	1.05	0.82	0.78
45	0.55	0.50	0.47	0.73	0.73	0.56
	WITH SUPERPLASTICIZER					
35	0.95	0.86	0.72	1.16	0.78	0.87
45	0.69	0.63	0.45	0.73	0.67	0.56

TABLE 9 — *Maturity efficiency* η_m *for 7 and 28 day strengths equivalent to PC concrete for series B, C, and D of Ref 13.*

Age Days	Reference Strength, MPa =	25	30	35	40
7	η_m for $r = 0.75$	0.69	0.68	0.67	0.67
	$r = 0.55$	0.63	0.59	0.65	0.63
	$r = 0.35$	0.59	0.58	0.56	0.55
28	η_m for $r = 0.75$	0.92	0.92	0.90	0.89
	$r = 0.55$	0.84	0.79	0.87	0.84
	$r = 0.35$	0.79	0.78	0.75	0.74

Maturity

Because of slow hardening, mixes containing alternative materials show improved strength-mass efficiency at greater ages. This is illustrated in Table 8.

Assuming M_B equal to that necessary for PC concrete to reach the reference strength, the data of Ref *13* was used to calculate the efficiency factors η_m shown in Table 9.

Adaptation of the η_m-values to provide a time-temperature basis of comparison resulted in the values given in Table 10.

These data indicate that a $r = 0.35$ mix would need 86 days at 5°C or 25 days at 40°C to equal the 7 day at 23°C strength of a $r = 1$ mix having the same w_o.

TABLE 10—*Curing period in days at various temperatures required to match the strengths of PC, f$_c'$ = 30 MPa. The PC concrete was assumed to have cured at 23°C for 7 or 28 days.*

Group	Mix	Days to Attain 30 MPa at Various Temperatures			
		5°C	10°C	23°C	40°C
a	r = 0.75	21	16	10	6
	r = 0.55	33	25	15	10
	r = 0.35	86	64	39	25
b	r = 0.75	105	79	48	30
	r = 0.55	172	129	78	52
	r = 0.35	309	232	141	93

NOTE—For Group a; the values of w_o were equal to that required to give f_c' = 30 MPa for PC control (r = 1) cured at 23°C for 7 days. Group b was similarly based on the r = 1 mix having f_c' = 30 MPa at 28 days.

Hydraulicity Factor

Table 11 gives an example of the variation of hydraulicity factor with varying amounts of BFS and FA. The most noticeable feature of these data is the reduction of hydraulicity factor for BFS from 28 to 90 days and the opposite effect for FA. These specimens were subjected to simulated mass curing by being stored in sealed containers until tested. The permeability of mixes containing FA was also substantially lower than these containing BFS [12].

Data on Fly Ash from Munday et al [Ref 14]

Efficiency factors for FA calculated from data in Ref *14* are given in Table 12. These data indicate a decline in efficiency with increase in FA content and a strong dependence on fineness as measured by the mass fraction passing the 45 μm sieve.

TABLE 11—*Hydraulicity factors calculated from data of Ref 12 for various mixtures of PC and BFS and PC and FA.*

Materials	Mix	Hydraulicity Factor at Age	
		28 days	90 days
PC	0.75	0.77	0.66
BFS	0.65	0.71	0.51
	0.55	0.55	0.51
FA	0.75	0.79	1.02
	0.65	0.46	0.73
	0.55	0.35	0.61

TABLE 12—*Efficiency factors for PC/FA mixtures as derived from Munday et al* [14].

	RELATED TO STRENGTH AND MASS RATIO r	
r	28 Days Strength f'_c, MPa	Efficiency Factor, η
0.82	57.5	0.67
0.74	47.0	0.50
0.55	37.0	0.39
0.55	17.0	0.38
0.47	57.5	0.30

RELATED TO LOSS ON IGNITION AND FINENESS		
Fineness[a]	Ignition Loss, %	Efficiency Factor
0.97	2.4	0.66
0.88	2.3	0.58
0.70	5.1	0.39
0.77	4.1	0.38
0.74	9.5	0.08
0.54	7.4	0.08

[a]Fineness = 1 − (fraction retained on 45-μm sieve).
r = PC/(PC + FA).

Conclusion

At the practitioner level the potential consumer of blast-furnace slag and fly ash, whether in blended cements or combined at the concrete mixer, needs more than the standard strength characteristics in order to assess costs and the performance of his choice.

The mass-strength efficiency factor is useful in that it takes account of strength, workability, and characteristic variability.

The maturity efficiency factor characterizes blended cements in terms of their response to different curing regimes.

APPENDIX I

Mass and Absolute Volume Calculated from Nominal Mix Proportions and Density and Interpolations from Strength Data

Let the masses of PC, BFS, aggregate, and water be P', B', A', and W' for the nominal mix and P, B, A, and W for a measured density of ρ kg m^{-3}.

The absolute volume of mix ingredients per cubic metre is

$$V_{mix} = [\rho/(P' + B' + A' + W')][P'/g_1 + B'/g_2 + A'/g_3 + W']$$

where g_1, g_2, and g_3 are the specific gravities of PC, BFS, and aggregate.

The air void content

$$V_{air} = (1000 - V_{mix})10^{-3} \text{ m}^3$$

The volume of paste

$$V_p = (P/g_1 + B/g_2 + W + V_{air})$$

The cementing quality of the paste is determined by the volume ratio of cementious material to water + air in the paste. We define quality index Q_i as follows

$$Q_i = (P/g_1 + B/g_2)/(W + A)$$
$$= (G/g_4)/(W + A)$$
$$= g_4/w_o$$

where

$$g_4 = (P)(g_1)/(P + B) + (B)(g_2)/(P + B),$$
$$w_o = (W + A)/C, \text{ and}$$
$$C = P + B.$$

w_o may be regarded as an "effective" W/C ratio where the air content is treated as an equivalent volume of water.

The viscosity of the paste varies with Q_i, and it is to be expected that, for constant workability, V_p will vary with Q_i and therefore with w_o.

Thus, if it is possible to fit experimental data to functions relating w_o and V_p and w_o and strength it is possible to estimate mix proportions yielding a given strength.

Consider, as an example, Mix 17 taken from Malhotra [13] having $\rho = 2320$ kg m^{-3}.

	Mass, kg	Volume, m$^3 \times 10^{-3}$
P	196	60.9
B	160	55.2
A	1816	674.9
W	165	165
	2337	956

where

$V_{mix} = (2320)(956)/2337 = 949$ m$^3 \times 10^{-3}$,

$V_{air} = 1000 - 949 = 51$ m$^3 \times 10^{-3}$,

$V_{paste} = 194.57/322 + 58.84/2.30 + 163.80 + 51$,

$\quad = 330$ m$^3 \times 10^{-3}$,

$w_o = [(165)(2320)/(2337) + 51]/[(2320)(196 + 160)/(2337)]$, and

$\quad = 0.61$

From data of Ref 13 the following may be calculated for $r = 0.55$.

	(a) No Superplasticizer					(b) With Superplasticizer		
Mix ref	3	9	17	25	31	10	18	26
Slump, mm	25	70	60	75	50	190	205	205
V_p	430	365	330	295	281	371	338	299
w_o^+	0.32	0.48	0.61	0.73	0.88	0.53	0.65	0.75
f_c' [28]	51	40	34	27	21	37	36	26

$$(a) \quad V_p = 263.5(w_o)^{-0.434} \qquad R^2 = 0.99$$

$$(b) \quad V_p = 254.0(w_o)^{-0.610} \qquad R^2 = 0.97$$

for both (a) and (b)

$$w_o = 2.76 - 1.409 \log(f_c') \qquad R^2 = 0.96$$

Interpolated Values

For f_c'

	MPa	20	25	30	35	40	45
(a)	w_o	0.93	0.79	0.68	0.58	0.50	0.43
	P	119	144	170	202	237	276
	B	97	117	139	165	194	226
	W'	201	206	210	213	215	216
(b)	P	···	145	176	215	259	···
	B	···	118	144	176	212	···
	W'	···	208	218	227	237	···

For $\sigma = 1.05 f_c'$

	MPa	20	25	30	35	40	45
(a)	w_o	0.88	0.74	0.63	0.54	0.45	0.38
	P	127	155	185	218	264	312
	B	104	126	151	179	216	255
	W'	203	207	212	214	216	216
(b)	P	···	158	194	235	294	···
	B	···	129	159	193	240	···
	W'	···	212	222	231	240	···

NOTE— $W' = W + V_{air}$

Target strength $\sigma = f_c' + 1.4(s)$

where s = standard deviation for group.

In this example $\sigma = 1.05 f_c'$ for both pure PC, that is, $r = 1$, and mixed cements.

APPENDIX II

Computation of Hydraulicity Factor (Ref 8)

Suppose that it is required to assess the cementing quality of material X when combined with PC in the mass proportions r units of PC and $1 - r$ units of X. Suppose further that one reference mix is made with r mass units of PC and $1 - r$ mass units of crystalline quartz ground to cement fineness (I), and the other reference mix be made with 1.0 mass units of PC. It is assumed that the starting porosity is the same in all cases. The strengths at a particular age are as follows

	Strength
For pure PC	σ_p
For mixture PC/X	σ_X
For mixture PC/I	σ_i

The hydraulicity factor H is given by

$$H = \frac{\sigma_x - \sigma_i}{\sigma_p - \sigma_i}$$

If the material of interest is as good as PC the value of $H \geq 1$.

If the material of interest has no cementing value it will yield the same strength as the inert reference material I and $H = 0$.

If the material X is harmful, H is negative.

Example

Material	28 Day Strength, MPa		
	$r = 1$	$r = 0.65$	$r = 0.55$
PC	48.8	...	...
Inert quartz	38.5	34.9	28.7
BFS	46.4	44.7	39.7
FA	46.6	41.2	35.8
H for BFS	0.77	0.71	0.55
H for FA	0.79	0.46	0.35

APPENDIX III

Efficiency Factor η_m Based on Maturity and the Time Taken to Reach f'_c

Maturity is defined as the product: (average temperature of the concrete $+ D$) (age in hours). D is the datum temperature at which the rate of hardening is zero. For PC concrete $D \approx -10°C$. For concrete made with BFS the datum temperature is approximately $-10°C$ at temperatures near 20°C, but it is different, and as yet undetermined, at very high or very low temperatures [17].

The strength-maturity relationship [16] which is used widely takes the form

$$f_c' = \alpha + \beta \log(M/100)$$

where α and β = empirical constants.

In the present study two different approaches were taken:
1. The efficiency factor η_m was defined as

$$\eta_m = (\log(M/100) \text{ for pure PC})/(\log(M/100) \text{ for the mix})$$

In this case it is assumed that the (water + air)/(PC + BFS) is constant for all values of r.

For example, the (water + air)/PC ratios for a range of strengths may be determined thus from the regression equations of Table 4

28 day strength, MPa	25	30	35	40
w_o for $r = 1$	0.86	0.75	0.66	0.58

Choosing $f_c' = 30$ MPa at 28 days as a reference strength, $w_o = 0.75$ for $r = 1$. The values of f_c' corresponding to $r = 0.75$, 0.55 and 0.35 may be calculated from the regression equations of Table 2.

For $r = 0.75$, as an example, the values are as follows

Age (days)	7	28	91	365
$\log (M/100)$	1.74	2.35	2.86	3.46
f_c' for $r = 0.75$	22	29	33	38

From these data a regression equation can be developed of the following form

$$\log(M/100) = P(f_c') + Q$$

For $r = 0.75$, $P = 0.11$, $Q = 0.68$, and the coefficient of determination $R^2 = 0.99$. Values of $\log(M/100)$ which would give the same strength as the PC mix, $f_c' = 30$ MPa, are now computed

$$\text{for } r = 0.75 \quad \log(M/100) = 2.55$$

$$\text{for } r = 1.00 \quad \log(M/100) = 2.35$$

The maturity efficiency factor $\eta_m = 2.35/2.55 = 0.93$
2. Use of the efficiency factor to estimate the length of time necessary to attain the same strength as the mix with pure PC ($r = 1$).

The number of days needed to reach the strength of the PC varies linearly with the $\log(M/100)$ value. For example the number of days curing of a $r = 0.75$ mix at an average of 10°C to reach the strength of a $r = 1$ mix curing at 22°C for 28 days is given by

$$\log(M/100) = 2.346/0.92$$

$$M/100 = 35\,481$$

For the $r = 0.75$ mix, the age in days at 10°C = 35 481/20(24) = 79 days.
For the $r = 0.75$ mix, the age in days at 40°C = 35 481/50(24) = 30 days.
3. M is the area under the age temperature curve + 10 (time in hours), where temperature is measured in °C and age in hours after final set. Known temperature variations can be easily taken into account. Use of the Plowman relationship [16] should be restricted, for the time being, to a range of 5 to 40°C, for mixes containing BFS.

APPENDIX IV

Influence of Variability on Efficiency Factor

If we assume three standards of control for $f'_c = 30$ MPa at 28 days we calculate

Control	v, %	σ, MPa
Very good	11	$30 + 1.4(3.3) = 34.6$
Fair	17	$30 + 2.4(5.1) - 3.5 = 38.7$
Poor	26	$30 + 2.4(7.8) - 3.5 = 45.2$

Water:cement ratios w_o and volume of paste V_p follow from equation pertaining to Ref *10* as shown in Tables 2 and 4.

Control	$r = 1$		$r = 0.75$		$r = 0.5$		$r = 0.25$	
	w_o	V_p	w_o	V_p	w_o	V_p	w_o	V_p
Very good	0.71	252	0.71	249	0.63	254	0.63	244
Fair	0.65	257	0.66	254	0.58	258	0.43	280
Poor	0.58	264	0.59	261	0.51	265	0.35	298

The corresponding quantities of portland cement follow

Control	Portland cement kg m^{-3} for various r^a			
	$r = 1$	$r = 0.75$	$r = 0.5$	$r = 0.25$
Very good	247	182	133	63
Fair	268	195	142	93
Poor	296	216	155	109
Typically	P_c	P_1	P_2	P_3
Efficiency	1.0	$\dfrac{(P_c - P_1)(10.75)}{P_1(0.25)}$	$\dfrac{(P_c - P_2)}{P_1}$	$\dfrac{(P_c - P_3)(0.25)}{P_1(0.75)}$
Efficiency	1.0	$\dfrac{3(P_c - P_1)}{P_1}$	$\dfrac{P_c - P_2}{P_1}$	$\dfrac{P_c - P_3}{3P_1}$

$^a r$ = mass ratio of $\dfrac{\text{PC}}{\text{total cement}}$.

References

[1] Emery, J. J., *Extending Aggregate Resources, ASTM STP 774,* American Society for Testing and Materials, Philadelphia, 1982, pp. 95–118.

[2] Meyer, A., and Mills, R. H., "Portland Composite Cement," Research Paper presented at First International Conference on the Use of Fly Ash Silica Fume, Slag, and Other Mineral Byproducts in Concrete, Montebello, Quebec. August 1983.

[3] Elmhirst, N. R. and Mills, R. H., "Control of Alkali-aggregate Reaction with Cement Blast-Furnace Slag and Fly Ash," Paper No. 46, Cements Division Annual Meeting American Ceramic Society, Pittsburg, May 1984.

[4] Meyer, A., "Investigations on the Carbonation of Concrete," *Proceedings V,* International Symposium on Chemical Cement, Tokyo, 1968, pp. 394–401.

[5] Mather, Bryant, "Cements Users' Expectations with Regard to Blended Cements," ACI SP-79, American Concrete Institute Detroit, 1983, pp. 255–265.

[6] Stutterheim, N. in *Proceedings,* Fifth International Symposium on Chemical Cement, Tokyo, 1968, pp. 270–276.

[7] Neville, A. M., *Properties of Concrete,* Pitman, London, 1981, pp. 653–665.

[8] Lea and Desch, *The Chemistry of Cement and Concrete,* 2nd Ed., Edward Arnold, London, 1956, p. 407.

[9] Warriss, B., *Strength of Concrete Containing Secondary Cementing Materials,* Special publication SP 79-28, American Concrete Institute, Detroit 1983, pp. 5389–5557.

[10] Mills, R. H., "Significance of Latent Energy in Hydraulic Slags to the Cement and Concrete Industry." Symposium on Utilization of Steelplant Slags, Australian Institute MM., Woolongong, 1979.

[11] Mills, R. H., "Assay of Blast-Furnace Slag as Cement," *Proceedings,* International Symposium on Concrete Technology, Monterey, Mexico, March 1981.

[12] Mills, R. H., "The Permeability of Concrete for Reactor Containment Vessels," Research Report INFO-0111, Atomic Energy Control Board Ottawa, 1983. Also issued as Publication 84-01 by University of Toronto, Department of Civil Engineering.

[13] Malhotra, V. M. in *Strength and Durability Characteristics of Concrete Incorporating a Pelletized Blast-Furnace Slag,* Publication SP-79, American Concrete Institute, Detroit, 1983, pp. 891–921.

[14] Munday, J. G. L., Ong, L. T., and Dhir, R. K. in *Mix Proportioning of Concrete with PFA,* Publication SP-79, American Concrete Institute, Detroit, 1983, pp. 267–288.

[15] Wood, K., Managing Director, Slagment Ltd., South Africa, private communication, Johannesburg, 1979.

[16] Plowman, J. M., *Magazine of Concrete Research,* Vol. 8, No. 22, March 1956, pp. 13–22.

[17] Mills, R. H. "A Preliminary Investigation of the Engineering Properties of Concrete Incorporating High-Magnesia Slag," *Transactions,* SAICE, Vol. 8, No. 9, Sept. 1958.

Richard A. Helmuth, [1] *David A. Whiting,* [1] *Vladimir S. Dubovoy,* [1]
Fulvio J. Tang, [1] *and Hugh Love* [1]

Performance of Blended Cements Made with Controlled Particle Size Distributions

REFERENCE: Helmuth, R. A., Whiting, D. A., Dubovoy, V. S., Tang, F. J., and Love, H., **"Performance of Blended Cements Made with Controlled Particle Size Distributions,"** *Blended Cements, ASTM STP 897,* G. Frohnsdorff, Ed., American Society for Testing and Materials, Philadelphia, 1986, pp. 106–127.

ABSTRACT: Blended cements were prepared from two controlled particle size distribution (CPSD) cements (95% finer than 31 or 23 μm) and one normally ground cement (345 m^2/kg Blaine), using the same clinker and gypsum. Pastes made with these cements blended with Class F or Class C fly ashes or ground granulated slag, and lesser amounts of ground limestone or silica fume, were tested for flow, strength development, and drying shrinkage. Three CPSD blended cements (two with Class F fly ash and one with slag, all with ground limestone) were selected for mortar and concrete tests along with two normally ground blended cements as controls. All five of these cements yielded mortars at 0.485 water/cement (W/C) with equal flow (121 $\pm$ 2%). Mortar strengths were higher at all ages with the CPSD blends than the controls.

The CPSD fly ash blended cement concretes prepared at 0.50 W/C required higher cement contents for 50 to 100 mm slump than did concretes prepared with normally ground blended cement. CPSD blended cements had slightly better compressive and flexural strengths and sulfate resistance than did their normally ground counterparts. Drying shrinkage of the concretes was lower when paste contents were lower, as with the CPSD slag blended cement.

KEY WORDS: admixtures, blast furnace slag, blended cements, cement pastes, cements, chemical admixtures, compressive strength, concretes, energy conservation, fly ash, mineral admixtures, mortars, particle size distribution, portland cements, portland pozzolan cements, portland slag cements

Since portland cement (PC) manufacturing is an energy intensive process [*1*], the use of cements made by blending PC with powdered mineral admixtures is an important means of saving energy. However, blended cements (BCs) usually develop strength more slowly than comparable PCs [*2*], unless the water/cement (W/C) ratio of the blended cement concrete is significantly reduced [*3*]. For this

[1]Principal research consultant, senior research engineer, research engineer, senior research chemist, and associate petrographer, respectively, Portland Cement Association, Skokie, IL 60077.

reason, BCs may not be acceptable for many uses, and such concretes may be vulnerable to frost action, leaching, and sulfate attack before they become mature enough to ensure durability.

Controlled particle size distribution (CPSD) PCs also have considerable energy savings potential [4-6]. Energy savings are expected to be achieved with CPSD cements because of (1) reduction of the specific surface area required in finish grinding, and (2) reduction of the amount of cement required to produce concretes of equal performance. Performance is maintained or improved by (1) reduction of the percentage of coarsely ground (greater than 30 μm) cement, and (2) reduction of the percentage of excessively finely ground (less than 5 μm) cement. These controls result in a narrower particle size distribution than in normally ground PCs.

CPSD cements can be made to have very rapid early strength development without being ground to excessively high specific surface areas. They are, therefore, especially well suited for use in BCs made with pozzolanic fly ashes (FAs) and ground granulated blast-furnace slags (GGBFSs); the rapid hydration of the CPSD cements compensates for the relatively slow reactions of FAs and slags. This paper presents the results of an investigation of 45 such BCs made with two different CPSD cements, one normally ground PC, and six different powdered mineral admixtures, two water reducing admixtures, and one accelerating admixture [calcium chloride ($CaCl_2$)]. The trial formulations were performance tested as cement pastes. Five BCs (three CPSD and two normally ground) were selected for further testing in mortars and concretes. The concrete tests were done mainly to demonstrate comparable performance in concretes of the BCs made with CPSD and normally ground cements, rather than to establish the relative merits of the different cements.

The expected improved early strength development of the CPSD BCs was confirmed in these tests, as well as comparable performance with respect to drying shrinkage and resistance to sulfate attack. There are, however, some differences between the results indicated by cement paste, mortar, and concrete tests. The mortar results showed the CPSD BCs to be clearly superior to the normally ground blended cements (NGBCs). This clear superiority was not achieved in the concrete tests, probably because the concrete mixes were designed for maximum economy rather than good workability.

Procedures

Materials preparation and testing procedures are very briefly summarized here; they are more fully described in Ref 6.

Materials Preparation

Portland Cements — Three PCs, one with a particle size distribution representative of current commercial practice and two CPSD cements, were ground from the same commercially produced cement clinker and gypsum rock. The normally ground cement was a commercial product, except that it was blended with some added ground clinker to adjust its gypsum content to optimum. The CPSD cements were prepared in the laboratory. Optimum gypsum contents were determined from compressive strength tests of 25-mm (1-in.) cubes of pastes made at 0.4 and 0.5 W/C and cured one day. All three PCs had equal gypsum contents and chemical compositions, which were determined by X-ray fluorescence (XRF) and atomic absorption spectroscopy (AA); sulfur trioxide (SO_3) was also determined by the Leco furnace method. Specific surface areas were determined by ASTM Fineness of Portland Cement by Air Permeability Apparatus (C 204–81); particle size distributions were determined with a Micromeritics Sedi-Graph 5000 particle size analyzer.

Blended Cements — Forty-five different BCs were also produced using five different mineral admixtures.

BCs for use in cement paste tests were prepared by dry blending small batches of PCs with mineral admixtures. The two FAs were each preblended for uniformity but not otherwise processed. After blending for uniformity, 32 kg (70 lb) of the ground granulated slag were particle size classified in the Vortec C-13 classifier to yield about 14 kg (30 lb) of coarse fraction (95% coarser than 15 μm). This coarse fraction was considered as a separate mineral admixture. The condensed silica fume (SF), a blended composite of 30 daily ferrosilicon production specimens, was used as received. The limestone used in this work was an Iowa limestone, crushed and ball-milled in this laboratory.

Binary, ternary, and quaternary BCs were prepared to explore the effects of composition on performance of CPSD BCs. The first series of blends was designed to contain major amounts of the two FAs (ASTM Classes F and C), the GGBFS, and the coarse fraction of the ground slag. Lesser amounts of finely ground limestone and one SF were also included in some blends. In this series blends were prepared using 70% PC and a total of 30% mineral admixtures on a volume basis.

Based on the results obtained with the first series, three CPSD BCs were selected for more detailed study. A second series of 15 additional BCs, including controls made using normally ground cement, was prepared. The percentage of PC in these blends was varied from 60 to 100% by volume.

Based on the cement paste results obtained in the first and second series of BCs, larger batches of five selected BCs were prepared for use in the mortar and concrete tests.

Cement Paste Tests

Paste mixes were made of the BCs and of the PCs used as their main constituents. The PCs were prepared at W/C ratios of 0.4 and 0.5.

The BCs were prepared using partial substitutions of equal volumes of mineral admixtures for PC. The paste mixes were also made with equal volume substitutions of blended for PC relative to the control mixes at W/C = 0.5. Since the densities of the mineral admixtures were slightly less than those of the PCs, the water/solids weight ratios varied slightly.

Tests were also made to determine the effects of water-reducing admixtures and $CaCl_2$ on the properties of pastes made with the three CPSD BCs prepared for use in mortar and concrete tests. In these tests, pastes were mixed at a W/C (water to blended cement weight ratio) of 0.4, with and without $CaCl_2$ or one of the two water-reducing admixtures, one conventional water-reducing and retarding admixture (A) and one high-range, water-reducing admixture (B). $CaCl_2$ was used at a dose of 2% by weight of cement. Admixture A (calcium lignosulfonate type) had a recommended addition rate of 3.2 to 4.9 mL/kg (5 to 7.5 oz/100 lb) of cement. Typical dosages of Admixture B (a sulfonated naphthalene-formaldehyde condensate) range from about 4 to 14 mL/kg of cement. The water-reducing admixtures were both used at doses of 7.5 mL/kg of PC in the blend. In tests of the PCs, this dosage had been found to yield 14 to 16% water reduction with Admixture A without excessive reduction of 1-day strengths, and to yield 20% or more water reduction with Admixture B without any significant retardation of strength development.

Mixing of cement pastes was done with a cool-bottom Waring blender using circulating water at 15°C and a blender power input of 200 W. A mix schedule of 30 s mix, 3 min rest, 30 s mix was used. Final paste temperatures slightly below 25°C generally were obtained by this method. Paste flow was determined by mini-slump cone pat area measurements [7]. Time of set in these paste tests was defined as the time at which the penetration of a modified Vicat needle (2-mm diameter) was less than 10 mm in pastes mixed at W/C ratios used.

Cement paste 25-mm cubes were cast with vibration (to remove entrapped air) in gang molds and stored in a moist cabinet at 23°C for 23 h before stripping, lapping, and compressive strength tests at 1 day, and at later ages after continued moist curing. Compressive strengths of 25-mm cubes were measured in duplicate (or triplicate for the PCs). The average coefficient of variation of compressive strengths by this method was 3%.

Cement paste specimens cast as thin slabs (15 by 80 by 3.2 mm) were moist cured for one day in their molds, stripped, and stored in small vials with a little water until measured at 7 or 28 days prior to drying. Specimens were dried in air over gently agitated saturated salt solutions of $Mg(NO_3)_2 \cdot 6H_2O$ (53% relative humidity). Lengths were measured with a dial indicator comparator (±0.001 mm) after curing for 7 or 28 days and after 21 and 28 days of drying, after which these thin slabs had reached constant length (within 0.01%).

Mortar Tests

The mortar tests were conducted according to the procedures in ASTM Test for Compressive Strength of Hydraulic Cement Mortars (using 2-in. or 50-mm cube specimens) (C 109–80) using a fixed W/C of 0.485, rather than adjusting the water content to give a fixed mortar flow as is specified for cements other than

PCs. Flow values were measured and compressive strengths determined at 1, 7, 28, and 90 days of moist curing at 23°C.

Concrete Tests

Concretes were prepared from cements blended with 25% FA and 5% ground limestone (cements 119P, 137P, and 128P), and from cements blended with 25% BFS and 5% ground limestone (cements 119S and 128S). Concrete mixtures using river gravel from Eau Claire, WI (19-mm maximum size) as coarse aggregate and a predominately dolomitic sand from Elgin, IL, as fine aggregate (fineness modulus 2.95) were designed for a slump of 50 to 75 mm and a W/C of 0.5. All mixing was done in Lancaster, countercurrent type open pan mixers of 0.092 and 0.021 m^3 (1.5 and 0.75 ft^3) nominal capacities. The mix cycle was 3 min mix, 3 min rest, 2 min mix.

Trial batches were prepared at constant slump and W/C to optimize the proportion of fine to coarse aggregate to obtain minimum cement content mixtures.

Mechanical Properties Tests

Compressive strength specimens were cast in 75 by 150-mm (3 by 6-in.) cardboard cylinder molds, and flexural strength specimens were cast into 76 by 76 by 286-mm (3 by 3 by 11¼ in.) steel prism molds. All specimens were consolidated by hand rodding and moist cured in a fog room (100% relative humidity) at 23 ± 2°C (73 ± 3°F). Elastic modulus concrete test specimens were cast into 152 by 305-mm (6 by 12-in.) steel cylinder molds and moist cured for 28 days.

All reported strengths represent the average of three cylinders tested in uniaxial compression by ASTM Test for Compressive Strength of Cylindrical Concrete Specimens (C 39–81), three prisms tested in flexure under third-point loading, or two cylinders tested for elastic modulus using ASTM Test for Static Modulus of Elasticity and Poissons Ratio of Concrete in Compression (C 469–81).

Drying Shrinkage — ASTM C 157

Concretes were cast into 76 by 76 by 286-mm (3 by 3 by 11¼ in.) steel prism molds fitted with gage points for subsequent length measurements. All specimens were moist cured for 7 days, then transferred to a laboratory maintained at 23 ± 2°C and 50 ± 5% relative humidity. Length readings were taken using a dial gage comparator, at intervals up to 12 months. All values represent the average of three specimens.

Resistance to Sulfate Attack

Concretes were cast into steel prism molds (previously mentioned). All specimens were moist cured for 28 days, at which point sulfate resistance testing was initiated. Specimens were stored in 10% sodium sulfate (Na$_2$SO$_4$) solution at 21 to 27°C (70 to 80°F) for 16 h, then placed in a room maintained at 54 ± 2°C (130 ± 3°F) for 8 h. This cycle was repeated daily. Percentage changes in weight and length are relative to initial measurements made at the end of the curing period.

Results and Discussion

Materials Analyses

Chemical oxide analyses of the clinker used to make the three PCs and the mineral admixtures used to make the BCs are given in Table 1. The potential compound compositions of the clinker was calculated to be 54% tricalcium silicate, 22% dicalcium silicate, 9% tricalcium aluminate, and 10% ferrite solid solution. The gypsum used had a sulfur trioxide (SO_3) content of 46.5%. The three cements were each prepared with 5.2% gypsum, which, as noted under Materials Preparation, prior testing had shown to yield maximum early strengths.

The Class C FA had a much higher calcium oxide (CaO) content than the Class F FA and was cementitious as well as pozzolanic. The condensed SF was relatively pure amorphous silica. The ground Iowa limestone used here was about 95% pure calcium carbonate ($CaCO_3$).

By X-ray diffraction and microscopic analysis, the ground granulated slag used had been found to be about 95% glassy, that is, less than 5% in the form of crystalline phases (mainly akermanite). These observations imply that it was of excellent quality for use in blended cements.

Blaine specific surface (BSS) areas, densities, and particle size distributions of these materials are given in Table 2. Also included in this table are the physical data for the three PCs which were used as the primary components in the BCs. Cement 119 has a particle size distribution representative of current commercial production and is referred to here as the normally ground, or control cement.

TABLE 1 — *Oxide analyses of clinker and mineral admixtures. weight percent.*

Oxides	Cement Clinker	Class C Fly Ash	Class F Fly Ash	Silica Fume	Ground Limestone	Ground Granulated BFSlag
SiO_2	22.10	35.70	50.0	91.6	3.26	41.0
Al_2O_3	5.24	20.30	23.8	0.6	0.67	5.78
Fe_2O_3	3.14	5.76	10.6	4.1	0.34	1.54
CaO	64.81	17.32	3.04	0.8	53.27	36.0
MgO	1.95	4.26	0.84	0.4	0.34	12.1
SO_3	0.71	3.05	0.68	0.9	0.12	2.86
K_2O	0.85	0.84	2.25	1.1	0.01	0.46
Na_2O	0.16	6.50	0.27	0.2	0.02	0.31
Free CaO	···	0.63	0.72	···	···	···
Loss on ignition	0.79	0.36	3.17	2.6	42.12	+0.87 (gain)

TABLE 2—*Specific surface areas, densities, and particle size distribution data.*

Material	BSS, m²/kg	Density, g/cm³	Cumulative Mass Percent Finer Than Particle Size Indicated Size, μm							95% to 5% Size Range, μm
			100	50	20	10	5	2	1	
Portland cement 119[a]	345	3.15	99	94	60	37	20	8	3	54 to 1.5
Portland cement 128[b]	323	3.15	99	99	90	44	19	5	2	23 to 1.9
Portland cement 137[b]	313	3.15	99	98	71	37	17	5	2	31 to 1.9
Class C fly ash	418	2.67	99	94	78	55	27	6	1	54 to 1.9
Class F fly ash	434	2.47	89	87	64	43	21	5	1	100 to 2.1
Silica fume	5500	2.34	100	99	99	98	96	91	82	3.2 to 0.1
Ground limestone	810	2.71	99	96	82	66	44	21	9	46 to 0.7
Ground granulated BFS	449	2.94	98	90	64	44	27	11	4	74 to 1.1
Classified granulated BFS	112	2.94	86	40	11	5	4	2	1	118 to 15

[a] Normally ground portland cement used to make blended cements used as controls.
[b] CPSD portland cement.

Cements 128 and 137 are CPSD cements having 95% finer than 23 and 31 μm, respectively.

Tests of Cement Pastes

Portland Cements Pastes — The three PCs were tested in cement pastes for flow (mini-slump cone pat area) and strength development up to 28 days. Results are given in Table 3. Both CPSD cements flowed less, or had higher water requirements for flow, than the normally ground control cement 119. The strengths of both CPSD cements exceeded that of the normally ground cement at 7 days, the <31-μm cement having lower 1-day strength because of its relatively low specific surface area.

Blended Cements Pastes, First Series — Test results obtained with the first series of BCs, which contained 30% mineral admixture (by volume), are given in Tables 4 and 5. Table 4 shows blend compositions, water/solid (W/S) weight ratios, strengths, and drying shrinkage data obtained with blends containing the <23-μm Cement 128. Strength results are expressed as the percentage of the strength of the CPSD PC at the same age. All four major mineral admixtures, that is, the FAs and slags, improved the flow properties of fresh pastes made with this cement, but replacement of 30% of the PC with any of these mineral admixtures resulted in considerable reduction in strength. Strength reductions were greatest with the Class F FA and the coarse slag, and least with the Class C FA, indicating it to be cementitious as well as pozzolanic. The Class F FA blends gave low early strengths, but strengths improved with longer curing times. The granulated slag blends showed strength developments intermediate between the results obtained with Class C and Class F FAs.

Substitution of 2% SF as a minor component in these blends reduced the flow (that is, increased the water requirement for flow) in all cases. The effects on strength were not consistent, both positive and negative effects occurred at different ages in the same blends.

Five percent ground limestone substitutions only slightly reduced the flow, but improved strengths at all ages in all blends.

Fifteen percent slag substituted for FA had little effect on strength except for reductions of the 1-day strengths of the Class C FA blends. Quaternary blends with coarse slag and limestone significantly increased the strengths relative to the straight Class F FA blend, but not relative to any of the others.

Drying shrinkage of the blends increased with curing time and with the amount of SF or Class C FA in the paste. Minor substitutions of the fine granulated slag increased the drying shrinkage of the Class F FA blend, but not that of the Class C FA blend. Coarse slag substitutions tended to reduce drying shrinkage.

Results obtained with the <31-μm cement 137 are given in Table 5. They are less complete but show similar trends. Since the data in Table 4 usually showed beneficial effects with 5% ground limestone, all of these blends were made with limestone. Five percent limestone substituted in the Class F FA blend yielded

TABLE 3 — *Properties of cement pastes made with portland cements.*

Cement	Nominal Size Range, μm	W/C	Pat Areas, in.²	Compressive Strength, psi			
				1d	3d	7d	28d
119[a]	54 to 1.5	0.4	7.6	4125	6600	7600	11 475
119	54 to 1.5	0.5	25.5	2375	4750	5750	8 900
128[b]	23 to 1.9	0.4	3.2	4650	8550	10 050	12 300
128	23 to 1.9	0.5	12.9	2650	5500	6900	8 800
137[b]	31 to 1.9	0.4	4.5	3600	8000	8 550	10 750
137	31 to 1.9	0.5	17.9	1925	4250	6325	7 875

Metric conversions: 1 in.² = 645.16 mm².
                     1 psi = 0.006895 MPa.
[a]Normally ground portland cement.
[b]CPSD portland cement.

TABLE 4—*Effects of several mineral admixtures in blended cements (70% CPSD cement 128 and 30% admixtures) on properties of cement pastes.*

Mineral Blend No.	Blended Cement Compositions, volume percentages				Fresh Paste Properties		Compressive Strength, % of control			Drying Shrinkage, %	
	Major Admixture	%	Other Admixtures[b]	%	Water/Solids g/g	Pat Area, in.²	1d	7d	28d	7d Cure	28d Cure
None	None	0	None	0	0.50	13.6	100 (2350 psi)	100 (6575 psi)	100 (9150 psi)	0.31	0.33
F1	F	30	none	0	0.53	19.5	53.2	54.7	59.8	0.28	0.32
F2	F	28	SF	2	0.54	11.2	48.9	60.8	72.7	0.30	0.38
F3[a]	F	25	L	5	0.53	17.0	59.6	63.1	61.2	0.30	0.34
F4	F	15	S	15	0.52	20.3	54.2	54.7	67.5	0.30	0.38
F5	F	15	Sc	15	0.52	22.5	50.0	55.9	61.7	0.28	0.34
F6	F	15	Sc&L	10 5	0.52	19.8	66.0	67.3	73.2	0.27	0.33
C1	C	30	none	0	0.52	24.3	80.8	66.9	77.6	0.34	0.44
C2	C	28	SF	2	0.53	14.9	81.9	76.8	71.6	0.39	0.50
C3	C	25	L	5	0.52	21.9	84.0	83.6	81.1	0.36	0.39
C4	C	15	S	15	0.52	20.1	54.3	72.6	79.8	0.34	0.43
C5	C	15	Sc	15	0.52	24.4	71.3	66.2	76.8	0.29	0.41
C6	C	15	Sc&L	10 5	0.52	21.5	65.5	63.9	76.8	0.34	0.36
S1	S	30	none	0	0.51	18.7	61.7	70.7	66.4	0.33	0.41
S2	S	15	F	15	0.52	20.3	54.3	54.8	67.5	0.30	0.38
S3	S	15	C	15	0.52	20.1	54.3	72.6	79.8	0.34	0.43
S4[a]	S	15	Sc&L	10 5	0.51	20.5	62.8	63.1	76.2	0.29	0.33
Sc1	Sc	30	none	0	0.51	24.8	51.1	63.1	63.4	0.27	0.31
Sc2	Sc	28	SF	2	0.51	13.5	54.3	65.4	65.6	0.31	0.37
Sc3	Sc	25	L	5	0.51	21.9	64.9	73.8	67.8	0.29	0.31
Sc4	Sc	15	F	15	0.52	22.5	50.0	55.9	61.7	0.28	0.34
Sc5	Sc	15	C	15	0.52	24.4	71.3	66.2	76.8	0.29	0.41

[a]Selected for further work.
[b]F = Class F fly ash.
C = Class C fly ash.
SF = silica fume.
S = ground granulated blast furnace slag.
Sc = ground granulated blast furnace slag (>15 μm).
L = ground limestone.
[c]Metric conversion: 1 in.² = 645.16 mm².
1 psi = 0.006895 MPa.

TABLE 5—Effects of several mineral admixtures in BCs (70% CPSD cement 137 and 30% admixtures) on properties of cement pastes.

Mineral Blend No.	Blended Cement Compositions, volume percentages				Fresh Paste Properties		Compressive Strength, % of control			Drying Shrinkage, %	
	Major Admixture	%	Other Admixture[b]	%	Water/ Solids, g/g	Pat Area,[c] in.²	1d	7d	28d	7d Cure	28d Cure
None	None	0	None	0	0.50	21.4	100 (1975 psi)[c]	100 (6225 psi)	100 (9325 psi)	0.30	0.33
F3[a]	F	25	L	5	0.53	25.5	61.5	63.8	67.0	0.32	0.37
F6	F	15	Sc &L	10 5	0.52	27.1	59.5	56.6	64.9	0.26	0.29
C3	C	25	L	5	0.52	25.8	78.2	73.9	75.3	0.36	0.37
C6	C	15	Sc &L	10 5	0.52	23.5	74.7	74.7	62.2	0.27	0.32
Sc3	Sc	25	L	5	0.51	25.4	56.2	58.6	63.2	0.28	0.30
S4	S	15	Sc &L	10 5	0.51	22.5	64.6	62.2	71.8	0.26	0.30

[a]Selected for further work.
[b]F = Class F fly ash.
C = Class C fly ash.
SF = silica fume.
S = ground granulated blast furnace slag.
Sc = ground granulated blast furnace slag (>15 μm).
L = ground limestone.
[c]Metric conversions: 1 in.² = 645.16 mm².
1 psi = 0.006895 MPa.

higher relative strengths than with cement 128. Comparable Class C FA and slag blends usually showed lower relative strengths than with cement 128.

The particular Class C ash used here yielded blended cements with very good strengths, but with drying shrinkages substantially greater than those of the controls, even with 5% limestone. Because our main objective was to determine the efficacy of using CPSD cements in BCs, rather than characterization of FAs, the more conventional and widely used Class F FA was used in further tests. Blend 123-F3 consisting of 70% <23 μm cement, 25% Class F FA, and 5% limestone, was selected for further tests because of its improved early strengths.

Granulated slag was selected as the second generic type of mineral admixture. Blend 128-F3 consisting of 70% <23 μm cement, 25% Class F FA, and 5% limestone, was selected for further tests because of its improved early strengths. 28-day strengths were higher and the drying shrinkage was substantially less than with the fine slag alone.

It had originally been supposed that the <23-μm cement 128 would be the cement that would benefit most from the improved flow and compensate for the reduced early strengths of blended cements. However, the results in Tables 4 and 5 showed that this was not always true; the relative strengths of the blends of 25% Class F FA, 5% limestone, and the <31-μm Cement 137 were higher than those made with the <23-μm cement 128. Therefore, blend 137-F3 was selected as the third blended cement for use in further tests.

Blended Cement Pastes, Second Series — Fifteen variations of the three selected BCs were prepared for tests to determine the effects of variation of total mineral admixture content. At each mineral admixture content the ratios of the two or three mineral admixture volumes were the same as in the selected blends 128-F3, 128-S4, and 137-F3 (Table 6 footnote). Compositions of these cements and test results are given in Table 6. Results for fly ash blended cements (FABCs) are also shown in Fig. 1. At ages of 1, 7, and 28 days, both the BCs containing the normally ground cement 119 and those containing CPSD cements 128 and 137 showed progressive reductions in strength as mineral admixture contents increased from 0 to 40%. As expected, the CPSD BCs showed smaller reductions in 1-day strengths than the normally ground cement. At later ages the CPSD and NGBC showed only small strength differences. For two of the three BCs made with 20% pozzolanic FA, the 91-day strengths were approximately equal to those of the controls without FA. The 91-day strength obtained with blend 137-F3 (Table 6) is so low as to be questionable.

Data in Table 6 do not provide much basis for selecting any particular percentage of mineral admixture for optimum performance of these BCs. At ages up to 28 days, the strengths decrease almost linearly with mineral admixture content, as clearly shown for the FA blends in Fig. 1. Only at 91 days do the data begin to suggest that some intermediate percentage may yield a maximum strength. The strengths at 91 days are relatively insensitive to the mineral admixture contents for values less than 30%, but begin to decrease precipitously at higher mineral

TABLE 6—Effects of mineral admixture content on performance of BCs made with PCs 119, 128, and 137.

Cement	Volume %	Mineral Blend No.	Mineral Blend[c,d]	Volume %	Water/Solids, g/g	Pat Area,[e] in.²	Compressive Strength, psi				
							1d	7d	21d	28d	91d
119[a]	100		None	0	0.50	25.5	2375	5750	...	8900	10 250
119	80	F3	F + L	20	0.52	27.9	1470	5100	6900	7725	10 350
119	70	F3	F + L	30	0.53	30.6	1115	4050	6175	6350	10 150
119	60	F3	F + L	40	0.54	30.0	730	2900	4750	5600	8 700
128[b]	100		None	0	0.50	13.6	2350	6575	...	9150	9 800
128	80	F3	F + L	20	0.52	20.2	1900	4800	6600	7050	9 850
128	70	F3	F + L	30	0.53	22.0	1387	3700	5750	6400	9 100
128	60	F3	F + L	40	0.54	24.3	1125	3100	4550	5275	7 800
137[b]	100		None	0	0.50	21.4	1975	6225	...	9325	10 250
137	80	F3	F + L	20	0.52	26.8	1525	4950	6575	7100	9 400 (?)
137	70	F3	F + L	30	0.53	28.5	1187	3850	5725	6375	9 550
137	60	F3	F + L	40	0.54	27.5	875	3200	4650	5150	8 150
119[a]	80	S4	S + Sc + L	20	0.51	30.2	1625	5700	7400	7900	9 800
119	70	S4	S + Sc + L	30	0.51	31.1	1177	4700	7000	7325	9 600
119	60	S4	S + Sc + L	40	0.52	31.2	850	4100	6500	7075	9 500
128[b]	80	S4	S + Sc + L	20	0.51	16.7	1600	5300	7200	7400	9 400
128	70	S4	S + Sc + L	30	0.51	21.5	1395	4900	6450	6675	9 000
128	60	S4	S + Sc + L	40	0.51	23.6	1067	3900	5950	6150	8 400

[a]Normally ground portland cement.
[b]CSPD portland cement.
[c]F + L blends in the ratio F:L of 25:5.
[d]S + Sc + L blends in the ratios S:Sc:L of 15:10:5.
[d]F = Class F fly ash.
S = ground granulated blast furnace slag.
Sc = ground granulated blast furnace slag (>15 μm).
L = ground limestone.
[e]Metric conversions: 1 in.² = 645.16 mm².
1 psi = 0.006895 MPa.

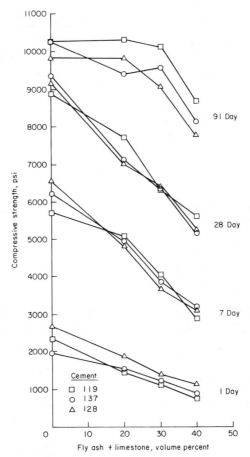

FIG. 1—*Paste strengths for FA blends.*

admixture contents. For these reasons, the BCs selected for mortar and concrete tests were those with 30% mineral admixture, Nos. 119-F3, 128-F3, 137-F3, 119-S4, and 128-S4. These three CPSD BCs and two controls were duplicated in much larger quantities, which were designated 119P, 128P, 137P, 119S, and 128S, the number denoting the portland cement used and the P and S indicating that the major blending ingredient was the pozzolanic (P) FA or the granulated slag (S). The compositions of these blended cements are given in Table 7.

Blended Cement Pastes with Chemical Admixtures —Results of tests of cement pastes mixed at water to solids mass ratio of 0.4 with and without three chemical admixtures are given in Table 8. The first three entries are for the three BC pastes without admixtures. Comparison of these results at 0.4 W/C with the results in Table 3 again shows that the strengths of the BC pastes are less than those of the corresponding PC pastes, except perhaps for cement 137P at 28 days, which is about 7% higher than that of cement 137.

TABLE 7—*Compositions of blended cements used for mortar and concrete tests.*

Blended Cement	Portland Cement, %		Major Admixtures, %[c]		Other Admixtures, %[c]	
			Blended Cement Compositions, volume percentages			
119P	119[a]	70	F	25	L	5
128P	128[b]	70	F	25	L	5
137P	137[b]	70	F	25	L	5
119S	119	70	S	15	Sc	10
					L	5
128S	128	70	S	15	Sc	10
					L	5

[a]Normally ground portland cement.
[b]CPSD portland cement.
[c]F = Class F fly ash.
L = ground limestone.
S = ground granulated blast-furnace slag.
Sc = ground granulated blast-furnace slag (>15 μm).

TABLE 8—*Effect of calcium chloride and two water reducing admixtures on properties of pastes made with 3 CPSD blended cements at a water to solids ratio of 0.4.*

Blended Cement	Admixture	Dose	Pat Areas,[c] in.[2]	Time of Set[d] h:min	Compressive Strength, psi			Modulus of Rupture, psi	
					1d	7d	28d	1d	28d
137P	None	⋯	7.4	5:24	2450	6900	11 550	790	2180
128P	None	⋯	5.3	5:38	2850	8750	10 725	790	2280
128S	None	⋯	5.5	5:03	2450	7200	11 000	820	2330
137P	CaCl$_2$	2[a]	6.5	2:12	3700	8450	11 600	820	1940
128P	CaCl$_2$	2[a]	6.3	2:30	4500	9450	11 675	960	1870
128S	CaCl$_2$	2[a]	5.7	2:02	3850	9250	12 950	830	1830
137P	A[e]	7.5[b]	20.6	>7:00 <23:00	240	7950	11 300	⋯	1970
128P	A	7.5[b]	17.8	>7:00 <23:00	250	7800	12 000	⋯	2200
128S	A	7.5[b]	17.8	>7:00 <23:00	60	8350	11 650	⋯	2520
137P	B	7.5[b]	25.5	>7:00 <23:00	2500	7700	9 550	840	2340
128P	B	7.5[b]	23.6	>7:00 <23:00	2700	8250	11 675	950	2350
128S	B	7.5[b]	26.2	>7:00 <23:00	2450	8300	11 800	860	2250

[a]Weight percent relative to portland cement fraction.
[b]mL/kg of blended cement.
[c]Metric conversions: 1 in.^{2} = 645.16 mm^2.
    1 psi = 0.006895 MPa.
[d]Nonstandard test.
[e]A = a calcium lignosulfonate type.
B = a sulfonated napthalene-formaldehyde condensate type.

The results for pastes with admixtures are consistent with our expectations, showing:

1. More rapid strength development with the 128P (23-μm CPSD cements) than 137P (31-μm CPSD cements).

2. Water-reducing admixtures to be almost as effective in improvement of paste flow with the 128 (23-μm) BCs as with the 137 (31-μm) BCs, which was not the case with the PCs.

3. Normal acceleration with calcium chloride.

4. Retardation of set and 1-day strength with admixture A.

5. Admixture B retarded set, but did not reduce strengths at one or seven days.

Mortar Test Results

Compressive strength test results of ASTM Method C 109-80 mortars made with the three PCs at a fixed W/C of 0.485 are given in Table 9. Also given are the results for the BC mortar tests; these were conducted using the same water to BC mass ratio of 0.485 as for the PCs.

Flow values of the five BCs ranged from 120 to 122.5%, which indicates that there were negligible differences between the water requirements of the different cements, and that the W/C requirements for adequate flow (110 ± 5%) were all less than 0.485; values for the PC mortars ranged from 0.46 to 0.47.

The CPSD BCs gave higher compressive strengths at all ages than the control cements made with the same mineral admixtures. Even the <31-μm cement with FA (137P) showed higher 1-day strengths than the control (119P) in these mortar tests.

TABLE 9—*Mortar tests of portland and BCs by ASTM C 109 except for the use of 0.485 water to blended cement ratio.*

Cement No.	Mortar Flow, %	Compressive Strength, psi[e]			
		1d	7d	28d	90d
119[a]	109[c]	1810	4330	5700	6420
137[b]	110[d]	1540	4690	6080	7000
128[b]	106[d]	2230	5150	6410	7220
119P	122.5	1380	3640	5480	7133
137P	122	1580	3880	5941	7525
128P	122.5	2020	4480	6360	7475
119S	120	1240	3380	5190	6525
128S	122	1340	3900	5570	6725

[a]Normally ground portland cement.
[b]CPSD portland cement.
[c]At W/C = 0.47.
[d]At W/C = 0.46.
[e]Metric conversion: 1 psi = 0.006895 MPa.

Concrete Results

Proportioning of the Concretes — Trial concrete mixtures designed for a slump of 50 to 75 mm at a W/C of 0.5 were tested to determine optimum proportions. Optimization curves for minimum cement content were determined for each of the five BCs, but it was found that the minimum cement content mixtures were lacking in cohesiveness and that the slump test was insensitive to relatively large increases in cement paste content, especially for cements 137P and 128P. Finally, it was decided to prepare these mixtures at somewhat higher cement contents in an effort to increase workability and cohesiveness of the concretes.

Concrete mix characteristics are shown in Table 10 for all five cements. The CPSD BCs tend to require higher cement contents or have lower slump values than the BCs made with normally ground cements and the same mineral admixtures. These results do not agree with results in Table 9, which showed no significant differences in flow of the mortars.

Mechanical Properties — Compressive and flexural strength data for these BC concretes are given in Table 11. Both of the FA CPSD BCs (137P and 128P) show slightly better strength development than the NGBC (119P) after three days of curing. For the slag BCs the rates of strength development are nearly equal. Also shown are the elastic moduli in compression results, which are also slightly higher for the CPSD FA BCs, than for the NGBC. These results are generally consistent with the more rapid hydration of the CPSD cements after one day of curing.

Drying Shrinkage — Drying shrinkage results for the BC concretes are given in Table 12. The shrinkage is slightly greater at each age for the CPSD FA concretes than the NGFAC, probably because of the greater degree of hydration of the CPSD BCs at the age of test, and because of the higher paste contents in these concretes. The CPSD slag cement, which had a lower cement content and a coarse slag fraction, showed less shrinkage than the NGSC.

TABLE 10 — *Concrete mix characteristics.*

| | | Batch Quantities,[a] lb/yd³ | | | | | | |
| | | | | Aggregate | | | | |
Cement	W/C	Cement	Water	Fine	Coarse	Fines, %	Slump, in.[b]	Air, %
119P[a]	0.50	499	250	1248	2100	37.5	2.9	1.0
137P	0.50	526	262	1214	2072	37.5	2.4	1.0
128P	0.50	536	267	1224	2012	37.5	4.0	0.8
119S[a]	0.50	508	254	1367	1957	41.0	3.3	1.5
128S	0.50	499	249	1288	2053	39.0	2.0	1.6

[a]Blended cement made with normally ground portland cement.
[b]Metric conversions: 1 lb/yd³ = 0.5933 kg/m³.
        1 in. = 25.4 mm.

TABLE 11 — *Mechanical properties of concrete made
with blended cements and moist cured at 23°C (73°F)*

Age at Test, days	Compressive Strength, psi[b]						Flexural Strength, psi[b]			Elastic Modulus, 10^6 psi[b]
	1	3	7	28	90	365	3	7	28	28
Cement										
119P[a]	1030	2090	3040	4920	6160	7130	450	600	740	4.2
137P	870	2390	3550	4990	6330	6710	480	650	750	4.3
128P	1040	2530	3680	5150	5950	7200	500	700	860	4.5
119S[a]	850	2070	3160	5190	5630	7030	440	720	870	4.5
128S	890	2310	3310	4980	5840	6760	500	680	870	4.5

[a]Blended cement made with normally ground portland cement.
[b]Metric conversion: 1 psi=0.006895 MPa.

Resistance to Sulfate Attack — Results of tests of the BC concretes for resistance to sulfate attack in 10% Na_2SO_4 solutions are shown in Figs. 2 and 3. Weight and length changes for the FA BCs are shown in Fig. 2 and for the slag BCs in Fig. 3. All BC concretes showed greatly increased resistance to sulfate attack relative to the concretes made with the corresponding PCs [6], and the effects of cement particle size distribution were clearly much less significant than the effects of the mineral admixtures. Nevertheless, Fig. 2 shows that the CPSD cement 137P showed much less rapid weight loss and less rapid expansion between 9 and 12 months than either the NGFAC 119P or the <23-μm CPSD cement 128P. The slag blended cements of Fig. 3 showed either rapid weight loss (128S) or rapid expansion (119S).

Comparison of Cement Paste, Mortar, and Concrete Results

Studies of the properties of blended cements made with NGPCs and CPSD PCs of identical compositions and equal amounts of the same powdered mineral admixtures yielded somewhat different results in cement pastes, mortars, and concretes.

TABLE 12 — *Drying shrinkage of BC concretes.*

Cement	Mineral Admixture	Drying Time: 1 wk	4 wk	2 mo	4 mo	8 mo	12 mo
		Drying Shrinkage, %					
119P[a]	fly ash	0.018	0.035	0.043	0.054	0.058	0.063
137P	fly ash	0.018	0.039	0.047	0.056	0.061	0.065
128P	fly ash	0.020	0.041	0.049	0.060	0.064	0.068
119S[a]	slag	0.018	0.038	0.045	0.058	0.065	0.066
128S	slag	0.012	0.029	0.040	0.050	0.059	0.059

[a]Blended cement made with normally ground portland cement.

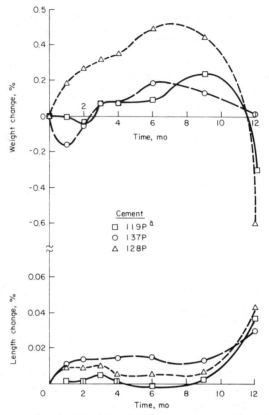

FIG. 2—*Length and weight change for FABC concretes in sulfate solutions.*

Studies of these cements in cement pastes have shown that:

1. The much higher water requirement for flow of CPSD PCs relative to NGPCs was greatly mitigated in some BCs so that CPSD BCs (with 25% Class F FA or slag and 5% ground limestone) flowed nearly as well as those made with NGBC pastes.

2. Water reducing admixtures were almost as effective in improving the paste flow of these blended cements made with a 23-μm maximum particle size cement as with a 31-μm cement.

3. The CPSD BC pastes had higher 1-day strengths than those made with normally ground cement.

4. Later age strengths differed little, with normally ground cements yielding slightly higher strengths at 91 days than the CPSD blended cements.

5. Drying shrinkage was significantly higher in some blends containing a Class C FA, finely ground granulated slag, or 2% substitution of condensed SF; substitution of 5% ground limestone for Class C FA reduced drying shrinkage, as did replacement with a coarser fraction of ground granulated slag.

6. Calcium chloride caused normal acceleration of set and strength devel-

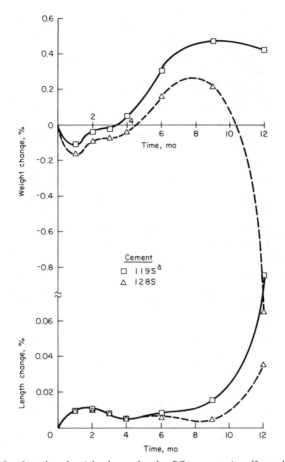

FIG. 3—*Length and weight change for slag BC concretes in sulfate solutions.*

opment in both types of blended cements, and retardation effects were also similar with a water-reducing, set retarding admixture.

Studies of the properties of mortars made with selected blended cements have shown that:

1. There were negligible differences between the flow values of mortars made with the different cements.

2. The CPSD blended cements yielded higher compressive strengths at all ages to 90 days than the control (normally ground) BCs made with the same admixtures.

3. The CPSD BCs often yielded higher strengths than even the corresponding PCs.

Studies of the properties of concretes made with the same selected blended cements have shown that:

1. The CPSD FA blended cements required concretes with higher cement contents than the corresponding NGBCs when designed for low slump and equal

W/C ratio; there was little difference in the cement contents of the concretes made with the slag blended cements.

2. Both FA CPSD BCs had slightly better strength development after 3 days curing than the NGBC.

3. Drying shrinkage of the CPSD FA BC concretes was higher than that of the concretes made with the corresponding normally ground cements; for the CPSD slag BC concretes the reverse was true.

4. The concrete showing the best resistance to sulfate attack in 10% Na_2SO_4 solutions was made with CPSD FA BC.

In some respects the results of the concrete tests appear to be in better agreement with the paste test results than with the mortar test results. The mortar results show better flow and strength development for the CPSD BCs relative to the normally ground cements than either the paste or concrete tests.

Although the pastes were mixed at slightly higher W/C ratios than the mortars (0.5 versus 0.485), their strengths were always higher than those of the mortars, especially after longer curing times. This is thought to result mainly from higher air contents (up to 12% by volume), which are characteristic of ASTM Method C 109-80 mortars. Knab, Clifton, and Ings [8] have suggested that the weakness of mortars (and presumably, concretes) relative to pastes is also a result of poor bonding between the cement paste and the standard quartz sand.

The data in Table 9 show that mortars made with the FA BCs containing 30% mineral admixtures gave 28-day strengths only slightly less than the corresponding mortars made with the unadulterated PCs. The 23-μm CPSD BC with FA (128P) actually gave mortar strengths even higher than those of the NGPC (119) without FA at all ages tested. This was also true, at later ages, for BCs 137P and 128S. Such results could not have been predicted from the paste test data shown in Table 6. These facts suggest that the air contents of the BC mortars were much less than those of the PC mortars. Although the air contents of the fresh mortars were not measured, this interpretation is consistent with the concrete mix characteristics. The air contents of concretes made with the BCs were only about half of those in the corresponding PC concretes.

Although the concrete strengths were all lower than the corresponding mortar strengths, the CPSD BC concretes results also showed better early strength development than those of the NGBC controls. These results would indicate positive energy savings could be achieved with CPSD BCs because of their reduced specific surface areas. However, the higher cement contents of the FA BC concretes resulted in slightly higher energy usage for these concretes. This resulted in "negative" energy savings estimated to be 3% of the BC manufacturing energy requirements [6].

This slightly negative result for the FA CPSD BC concretes versus the NGBC concretes may have been a result of the use of low slump, relatively low cement content concretes. In similar comparisons of the results obtained with normally ground and CPSD PC mortars and concretes, it was found that the strength of more workable (100 to 120-mm slump) concretes were in much better agreement with the mortar test results than those of very stiff mixes like those used to make these BC concretes. It seems probable that even more favorable résults would have been obtained with the CPSD BC concretes if they also had been designed for 100 to 120-mm slump.

Conclusions

Cement paste, mortar, and concretes made with CPSD blended cements have been shown to have properties approximately equal or superior to those made with normally ground blended cements of the same compositions. The major benefit in the use of controlled particle size distribution blended cements may be to produce concretes with properties, especially at early ages, which are comparable to those obtained with portland cements. In this way, blended cements may become more readily acceptable to more users. Since the use of blended cements is in itself a major energy savings measure, wider use of higher early strength blended cements could contribute to energy conservation in the United States.

Acknowledgments

Funding for this research was provided by the U.S. Department of Energy under Cooperative Agreement DE-FC07-81-CS40419 with the Portland Cement Association's Construction Technology Laboratories. Cement, cement clinker, and gypsum rock were provided by the National Gypsum Company, Cement Division, Alpena, Michigan. Fly ashes were supplied by American Fly Ash Co., Des Plaines, Illinois. The ground granulated slag was provided by Standard Slag Cement Co., Fruitland, Ontario, Canada. Silica fume was provided by American Admixtures and Chemicals Corp., Chicago, Illinois. Admixture A (WRDA 79) was provided by W. R. Grace & Co., Pompano Beach, Florida. Admixture B (Mighty 150) was provided by ICI Americas, Inc., Specialty Chemical Division, Wilmington, Delaware.

References

[1] "Energy Conservation Potential in the Cement Industry," Conservation Paper Number 26, Federal Energy Administration, Washington, DC, 1975; NTIS PB 245 159/AS.
[2] Klieger, P. and Perenchio, W. F., "Laboratory Studies of Blended Cements: Portland Pozzolan Cements," Bulletin RD013, Portland Cement Association, R&D, 1972.
[3] Berry, E. E. and Malhotra, V. M. in *Proceedings,* American Concrete Institute, Vol. 77, No. 2, 1980, pp. 59–73.
[4] Helmuth, R. A., "Energy Conservation Potential of Portland Cement Particle Size Distribution Control, Phase I. Final Report," by Construction Technology Laboratories for the U.S. Department of Energy, ERDA Contract No. EC-77-C-02-4269, March 1979; NTIS COO-4269-1.
[5] Helmuth, R. A., "Improved Cement and Energy Savings with Particle Size Control," *Proceedings,* IEEE Cement Industry Technical Conference, Lancaster, PA, 10–14 May 1981.
[6] Helmuth, R. A., Whiting, D. A., and Gartner, E. M., "Energy Conservation Potential of Portland Cement Particle Size Distribution Control, Phase II, Final Report," by Construction Technology Laboratories for the U.S. Department of Energy, U.S. DOE Cooperative Agreement DE-FC07-81CS40419, July 1984.
[7] Kantro, D. L., *Cement, Concrete, and Aggregates,* Vol. 2, No. 2, Winter 1980, pp. 95–102.
[8] Knab, L. I., Clifton, J. R., and Ings, J. B., *Cement and Concrete Research,* Vol. 13, 1983, pp. 383–390.

R. Douglas Hooton [1]

Permeability and Pore Structure of Cement Pastes Containing Fly Ash, Slag, and Silica Fume

REFERENCE: Hooton, R. D., "**Permeability and Pore Structure of Cement Pastes Containing Fly Ash, Slag, and Silica Fume,**" *Blended Cements, ASTM STP 897,* G. Frohnsdorff, Ed., American Society for Testing and Materials, Philadelphia, 1986, pp. 128–143.

ABSTRACT: As part of research to develop a highly durable concrete container for radioactive waste disposal in chloride and sulfate bearing granite groundwaters, a variety of cement pastes were studied. A sulfate resisting portland cement was used with various replacement levels of Class F fly ash and pelletized blast furnace slag at a water to solids ratio (W/S) = 0.36. Blends with fly ash, slag, and silica fume were also combined with a super water reducer at W/S = 0.25. Results are presented for strength development, permeability to water, and pore size distribution after 7, 28, 91, and 182 days moist curing. As a direct measure of durability, after 91 days moist curing, paste prisms were immersed in both de-ionized water and a synthetic chloride and sulfate bearing groundwater at 70°C.

While all three supplementary cementing materials (mineral admixtures) reduced ultimate permeabilities, silica fume was more effective in reducing permeability at early ages. Silica fume was also the most effective in reducing calcium hydroxide contents of the pastes while slag was the least effective; only reducing calcium hydroxide levels by dilution of the portland cement. From preliminary analysis, there does not appear to be a way of accurately predicting permeability from porosity or pore size parameters alone.

KEY WORDS: cement paste, slag, fly ash, silica fume, permeability, mercury porosimetry

The use of blended cements or supplementary cementing materials (mineral admixtures) has been reported to increase the resistance of concrete to deterioration by aggressive chemicals such as chlorides [1,2] and sulfates [3–5], to deleterious alkali-aggregate expansions [6,7] and to corrosion of reinforcement [8]. Much of this increase in durability has been attributed to reduced permeability (and reduced ion diffusion) resulting from a finer pore structure [1,5] and to reduced contents of easily leached or reacted calcium hydroxide [9] in the hardened paste fraction.

---

[1] Concrete materials engineer, Civil Research Department, Ontario Hydro Toronto, Ont. Canada.

As portland cement hydrates, large capillary pore spaces are filled in with hydration products, thus refining the size of these large pores and at the same time increasing the volume of very fine gel pores or interlayer pores.

Supplementary cementing materials allow further refinement of the pore structure by reacting with the calcium hydroxide (CH) liberated in the hydration of the C_3S and C_2S^2 in the portland cement, thus forming additional or secondary cementitious products (CSH) within the framework of the hardened cement paste.

In addition to refining pore sizes, the volume previously occupied by the CH consumed in the secondary reactions is also filled in.

Until recently, there has been very little data reported on the effects of supplementary cementing materials on permeability, largely due to a general lack of permeability testing or standard methods of test [10–12].

More information on the refinement of pore structure by supplementary cementing materials has been reported [1,5,10] due to the relative ease of obtaining pore size distributions using mercury intrusion techniques.

It is assumed that there is a relationship between permeability and pore size distribution, and this has been investigated [10, 13, 14], but only the work by Manmohan and Mehta [10] has specifically dealt with blended cements. However, Feldman [15] has shown that the pore size distributions of blended cements are likely coarser than indicated by mercury intrusion, and structural damage is caused by the procedure.

Also, the role of supplementary cementing materials in reducing the CH contents of hardened pastes was found to be at least as important as the refinement of pore size distributions for increased sulfate [5] and chloride [1] resistance.

Given this background, the primary objective of the present study is to investigate potential improvements to the permeability, pore structure, and chloride/sulfate durability of a sulfate resistant portland cement offered by partial replacements with locally available fly ash, slag, and silica fume. Secondary objectives are to investigate the possibility of predicting permeability from mercury intrusion pore size distributions and predicting chloride/sulfate durability from variations in permeability, pore size distributions and CH.

Procedures

Materials

A sulfate resisting portland cement (SRPC) conforming to ASTM Type V, CSA Type 50 and API Class G requirements was used along with partial replacements (by volume) by ASTM Class F fly ash (FA), pelletized blast furnace slag (BFS) and condensed silica fume (SF). The properties of these materials are given in Table 1.

---

[2]Cement chemical notation: C = CaO, S = SiO_2, A = Al_2O_3, F = Fe_2O_3, H = H_2O.

TABLE 1 — *Properties of cementing materials.*

Cementing Materials	Sulphate Resistant Cement,[a] SRPC	Class F Fly Ash, FA	Pelletized Blast Furnace Slag, BFS	Condensed Silica Fume, SF
Oxides, %				
CaO	63.6	2.8	38.16	0.17
SiO$_2$	22.9	42.3	37.89	92.7
Al$_2$O$_3$	2.80	23.2	8.39	0.17
MgO	3.31	0.96	11.55	0.59
Fe$_2$O$_3$	3.64	14.7	(1.42 AS Fe)	0.29
SO$_3$	2.01	0.87	(2.01 AS S)	0.36
K$_2$O	0.45	2.1	0.47	0.95
Na$_2$O	0.47	0.66	0.17	0.16
P$_2$O$_5$	0.06	0.49	<0.01	0.06
LOI	0.68	7.96	⋯	3.24
Relative density	3.17	2.31	2.94	2.22
Blaine, (m^2/kg)	377	381	410	6000[b]

[a] Bogue composition: C$_3$S = 53.8%, C$_2$S = 25.1%, C$_3$A = 1.8%, C$_4$AF = 11.1%.
[b] Higher surface areas obtained by N$_2$ adsorption are more accurate.

By X-ray diffraction, the FA contained glass (amorphous halo maximum at 24°/2θ, Cukα), with substantial quantities of mullite, quartz, and magnetite. It is interesting to note that by Diamond's method [16], for 2.8% calcium oxide (CaO), the maximum of the amorphous halo should be 23.4°/2θ ± 0.4% (approximately). The BFS was mainly glass (amorphous halo maximum at 30.5°/2θ) with small quantities of crystalline melilite and merwinite. The SF was mainly glass (halo maximum at 21.3°/2θ) with a trace of crystobalite and an unidentified peak at 2.52 Å.

De-ionized water, and in some cases a melamine formaldehyde condensate-based super water reducer (SWR) having 20% solids, were used to obtain workable pastes. SRPC was used as the base cement to provide resistance to the sulfates contained in granitic groundwaters (the Canadian reference concept for radioactive wastes is geologic disposal in granite). However, these granitic groundwaters also contain high concentrations of chlorides and as Feldman found [9], SRPC can be less resistant than normal portland cement to concentrated chloride solutions due to its higher C$_3$S and C$_2$S contents which result in larger quantities of CH in the hardened paste. CH is readily leached, thus opening the interior of the paste to chloride attack. Therefore, the supplementary cementing materials were employed in quantities thought to provide improvements to durability by reduction of CH in the hardened paste (25 and 35% by volume FA, 50 and 65% BFS, 10 and 20% SF). Because of the effect of its extreme fineness on workability, the SF was only used in conjunction with the SWR. For comparison, the SRPC paste and the higher replacement level with FA and BFS were also made with the SWR admixture.

Specimen Preparation

During initial attempts to mix pastes to a constant workability, both the flow cone [17] and the minislump [18] procedures proved to be unsatisfactory for the pastes containing SF. Therefore, somewhat arbitrarily chosen water to solids ratios (W/S) of 0.25 and 0.36 were employed to produce workable pastes with and without use of the SWR admixture, respectively. The water and solids fractions of the SWR admixture were included in calculations of W/S.

Pastes were mixed under vacuum in a high-speed, stainless steel, 3.8 L capacity blender for three 1-min intervals separated by 30-s rest periods (to minimize heat generation). Twelve, 50-mm cubes were cast for compressive strength determination after 7, 28, 91, and 182 days moist curing.

Six, 25 by 25 by 125-mm gage length prisms were cast for length and weight change determinations during moist curing (to 91 days) and subsequent exposure to both de-ionized water and a simulated chloride/sulfate bearing groundwater at 70°C (35 000 mg/L Cl^-, 770 mg/L SO_4^{2+}, Ca^{2+} = 14 000 mg/L, Na^+ = 5000 mg/L, pH = 6.8).

Specimens for permeability and pore size determinations were obtained from paste cast into a 100 by 200-mm metal cylinder mold. To avoid bleeding and segregation before setting, the top of the cylinder was sealed with a rubber gasket and clamped between two circular end plates. This assembly was then placed on a roller mill and rotated about the cylinder axis for 18 to 21 h at 46 rpm. All molds were stripped at 24 h, and specimens were subsequently moist cured.

The cylinder was later sliced into three 40-mm disks and one 25-mm disk (with both end pieces discarded), three for permeability measurements and one (quartered) for pore size determinations. Initially, as a check on uniformity of the hardened paste from the rotated cylinders, Rockwell hardness tests were performed across one disk cross section and very little variation was observed.

Equipment

Cumulative pore size distributions were obtained by mercury intrusion using a Quantachome automatic scanning porosimeter with intrusion pressures up to 414 MPa (60 000 psi) (using a contact angle of 140°, pores down to 0.018 μm (18 Å) equivalent radius can be intruded). A microprocessor was obtained part way through the program which allowed plotting of differential pore size distributions (dV/dP). From earlier pore size data, maxima from the differential curves were simply estimated from the maximum slope of the cumulative intrusion curve. Specimens were dried for 16 to 20 h at 110°C, crushed to pass a 4.7 mm (No. 4) and be retained on a 2.4 mm (No. 8) sieve then dried again at 110°C prior to test.

Water permeabilities were measured using specimens and test cells similar to those described by MacInnis and Nathawad [19]. For each paste composition, sand filled epoxy rings were cast around the roughened perimeters of the three, 40 by 100-mm diameter disks. Each cell (Fig. 1) consisted of 2 end plates fitted

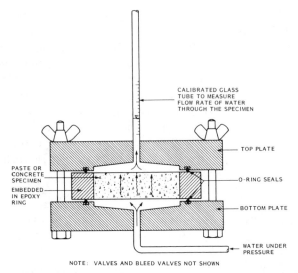

CALIBRATED GLASS
TUBE TO MEASURE
FLOW RATE OF WATER
THROUGH THE SPECIMEN

TOP PLATE

PASTE OR
CONCRETE
SPECIMEN
EMBEDDED
IN EPOXY
RING

O-RING SEALS

BOTTOM PLATE

WATER UNDER
PRESSURE

NOTE: VALVES AND BLEED VALVES NOT SHOWN

FIG. 1—*Sketch of permeability cell for specimens of up to 150 mm diameter.*

with O-rings to bear against the epoxy ring. The bottom faces of the specimens were exposed to de-aired, de-ionized water which was pressurized with a piston driven by nitrogen gas. The top faces of the specimens were saturated, and the flows of permeating water were measured in calibrated 10-mL glass tubes attached to the top plates of the cells (readings were taken to the nearest 0.01 mL).

Specimens were saturated in water under vacuum for 24 h prior to each test, and permeability tests were carried out for up to 24 h. Constant pressures of 172 kPa (25 psi) were applied for 7 day old specimens and 690 kPa (100 psi) at later ages. Flow volume-time curves were plotted, and, after initial, nonlinear portions were deleted, permeability constants were calculated using the Darcy equation re-arranged as follows

$$\text{permeability} = K(\text{m/s}) = (\Delta V/\Delta T)(t/P \cdot A))$$

where

$\Delta V/\Delta T$ = volume of water permeating over a given time (m^3/s),

t = specimen thickness (m),

A = cross-sectional area of specimen (m^2), and

P = pressure differential across the specimen (metre of water head).

Results

Workability

Flow cone times and mini-slump measurements are given in Table 2. It can be observed that there is little relationship between the two measures of workability.

TABLE 2 — *Workability of pastes.*

Cementing Materials	W/S	Flow Cone Time, s	Mini-Slump Area, cm²
100% SRPC	0.25ᵃ	22	143.9
35% FA	0.25	90	136.1
65% BFS	0.25	70	151.0
10% SF	0.25	15	76.8
20% SF	0.25	no flow	25.8
100% SRPC	0.36	13	53.5
25% FA	0.36	44	38.1
35% FA	0.36	35	32.9
50% BFS	0.36	26	34.8
65% BFS	0.36	37	30.3

[a] All mixes at W/S = 0.25, contained 3% super water reducer, except 5% for the 35% FA mix.

Neither method was suitable for the pastes containing SF which while very workable, exhibited thixotropic properties. It is intended to test duplicate pastes using a rotational viscosimeter at a later date.

Strength Development

Values for compressive strength development are given in Table 3. At W/S = 0.36, increasing replacements with FA or BSF resulted in larger reductions in compressive strength at 7 days. However, after 28 days both pastes containing slag attained similar strengths to the SRPC paste. The replacements by volume rather than weight likely accounted for the strengths of the blends not being higher.

As would be expected, with the super water reducer and W/S = 0.25, the strengths of the SRPC, slag, and FA pastes are significantly higher at all ages. The large strength gain of the FA paste between 28 and 91 days is consistent with its counterpart at W/S = 0.36, benefitting more from extended periods of moist curing than either the BFS or SF. However, it should be noted that for the FA

TABLE 3 — *Strength development of pastes.*

Cementing Materials	W/S	Compressive Strength, MPa			
		7 Days	28 Days	91 Days	182 Days
100% SRPC	0.25	82.0	99.7	109.0	112.8
35% FA	0.25	70.1	82.7	122.8	126.4
65% BFS	0.25	80.5	77.6	91.9	114.3
10% SF	0.25	71.9	75.8	86.6	88.7
20% SF	0.25	76.3	89.8	97.9	107.2
100% SRPC	0.36	52.9	66.6	76.9	80.2
25% FA	0.36	40.5	58.2	77.3	80.1
35% FA	0.36	36.0	59.0	80.2	82.9
50% BFS	0.36	38.2	70.7	80.3	85.1
65% BFS	0.36	33.6	64.3	75.2	80.8

TABLE 4 — *Coefficients of Permeability of pastes.*

Cementing Materials	W/S	Coefficient of Permeability, 10^{-13} m/s			
		7 Days	28 Days	91 Days	182 Days
100% SRPC	0.25	6.3	3.8	1.3	0.3
35% FA	0.25	5.1	1.3	0.5	<0.3
65% BFS	0.25	28.0	5.1	0.1	<0.1
10% SF	0.25	10.0	0.9	0.6	0.4
20% SF	0.25	6.3	<0.1	0.3	<0.1
100% SRPC	0.36	340.0	2.5	1.3	1.0
25% FA	0.36	140.0	2.5	0.5	0.3
35% FA	0.36	280.0	...	...	...
50% BFS	0.36	265.0	6.3	0.6	<0.3
65% BFS	0.36	2500.0	6.3	1.3	0.4

composition, 5% super water reducer (by weight of cementitious materials) was employed instead of 3% used in the other compositions.

At all ages, the paste containing 20% SF exhibited higher strengths than the one containing 10%. Again, these strengths would have likely exceeded those for the SRPC pastes if replacement by weight had been adopted instead of by volume.

Permeability

The permeability coefficients calculated for the pastes are given in Table 4. In some cases, no flows could be determined as indicated by the "less than" symbol. The average coefficient of variation for the permeability results was calculated to be 51.9%, based on 79 sets of 3 replicate tests. While this degree of variability might appear to be high, it may be typical of water permeability testing since in the only other published data giving coefficients of variation, values ranged from 44 to 148% (based on 17 sets of 4 replicate tests [20]).

For pastes made at W/S = 0.25, 20% SF replacement was the most effective in minimizing permeability at 28 days, although all of the supplementary cementing materials offered improvements to the permeability exhibited by the SRPC. After 91 and 182 days curing, the permeabilities of pastes with BSF and FA showed substantial improvements. The anomolous result for the 20% SF at 91 days is thought to be due to experimental sensitivity at extremely low flow rates close to the detection limit of the test procedure.

For pastes made at W/S = 0.36, FA and BSF replacements reduced permeabilities at ages of 28 days or more. After only 7 days curing, large replacements with both BFS and FA resulted in retarded development of physical properties as would be expected. Unfortunately, the permeability results beyond 7 days curing for the paste containing 35% FA were obviously in error, likely due to leakage at the paste-epoxy interface, and have not been reported.

It is difficult to compare the permeabilities obtained in this study with those found in the literature since the values obtained are sensitive to the applied pressure gradient [12,21], the magnitude of triaxial constraint and the direction of test relative to casting [11], air in the system, temperature, and boundary conditions [12].

It should be also noted that permeabilities obtained in laboratory studies, even for concrete specimens, cannot be simply translated to values in structures due to the presence of unavoidable defects in structures such as areas of incomplete compaction, thermal or moisture induced cracks, bleeding cavities, and fissures at reinforcement [11].

Pore Structure

The mercury intrusion porosity curves for each paste after 7, 28, 91, and 182 days moist curing are shown in Figs. 2 and 3. In addition, for ease of comparison, the volumes of pores larger than 0.025 μm radius (similar to the 500 Å diameter mentioned by Mehta [5]) are given in Table 5 and total intruded porosities are given in Table 6. From the figures, the decreasing size and volume of intruded porosity with curing period can be observed. Slight overlaps at fine pore radii in a few of the pastes with W/S = 0.25 are not considered significant and likely represent minor variations due to the small size of the specimens employed.

For W/S = 0.25, the pore structure of the pastes became finer with increasing quantities of SF. As well, with 20% SF, the total intruded porosity was significantly less. With 65% BFS, for up to 91 days curing, total porosities were higher than the SRPC, but mainly in pore sizes smaller than 0.025 μm radius. However, after 182 days curing, the porosity of the slag paste was much lower and of finer sizes, similar to the ones containing SF. The increasing slope of the fine end of the intrusion curves (that is, at maximum intrusion pressure) for the blended cement pastes indicates that there is considerable porosity that is not being intruded. The pore structure of the paste containing FA was not finer than that for the SRPC at any age which is inconsistent with the permeability and strength data.

For W/S = 0.36, total porosities were much larger, and, consequently, the reduction in porosities with increased periods of curing are more apparent. In relation to the SRPC paste, 25 and 35% FA replacements resulted in larger fractions of coarse size pores at 7 days, but the volume of large pores were similar by 28 days. With 35% FA replacement, the volume of pores larger than 0.025 μm radius was smaller at both 91 and 182 days. In all cases, the total intruded porosity of the pastes containing FA were higher than those of the SRPC paste. After 28 days curing, the higher porosities generally occurred in the pores finer than 0.005 μm radius, indicating a larger quantity of interlayer or gel pores (that is, more CSH).

With 50 and 65% replacements of BFS, large volumes of coarse sizes pores were observed after 7 days curing, but, at later ages, pore sizes were finer than the SRPC paste. The reduction in the larger, more permeable pores after 182 days curing was better with 50% BFS, than 65% BFS which is also reflected in its higher compressive strength and lower permeability.

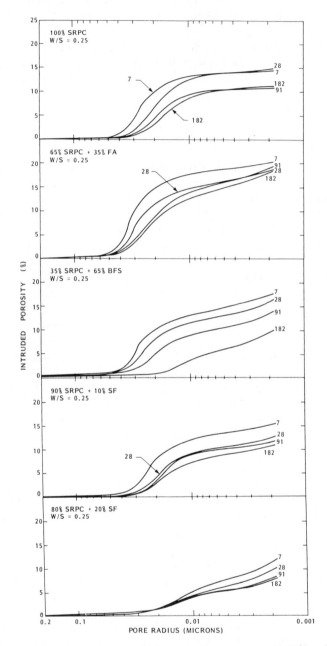

FIG. 2—*Cumulative pore size distribution curves for pastes cast with W/S = 0.25.*

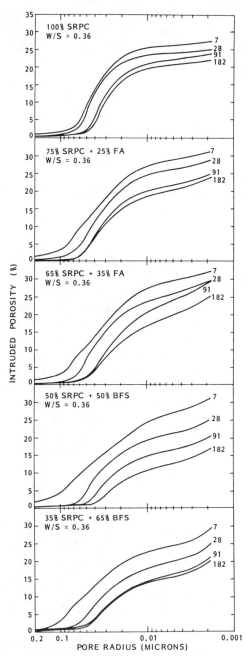

FIG. 3—*Cumulative pore size distribution curves for pastes cast with W/S = 0.36.*

TABLE 5 — *Volume of pores larger than 0.025 μm radius.*

Cementing Materials	W/S	Porosity Larger Than 0.025 μm Radius, %			
		7 Days	28 Days	91 Days	182 Days
100% SRPC	0.25	7.5	3.6	2.2	1.3
35% FA	0.25	12.6	9.2	6.1	5.2
65% BFS	0.25	8.0	5.4	2.7	0.6
10% SF	0.25	4.5	2.1	1.4	1.4
20% SF	0.25	0.9	0.9	1.1	1.1
100% SRPC	0.36	20.8	19.9	14.8	13.2
25% FA	0.36	20.3	17.8	14.0	12.5
35% FA	0.36	19.7	17.6	11.8	9.7
50% BFS	0.36	16.9	13.6	9.0	5.6
65% BFS	0.36	17.4	12.7	8.9	8.6

Hydration Products

The hardened pastes were examined by X-ray diffraction after 7, 28, 91, and 182 days curing. Amorphous CSH and unreacted SRPC (by alite peaks) were noted in each case with CH present in most cases. With the pastes containing SF, there was no evidence of the amorphous halo due to SF even after only 7 days curing. Using the SRPC paste as a reference, and correcting for the percentage of SRPC in the blended cement pastes, it was evident that with BFS or SF replacements, less alite had hydrated (that is, a slower reaction rate than for the SRPC alone) after 7, 28, and 91 days curing. However, for the paste containing FA, the reaction rate of alite was similar to that observed on the plain SRPC paste. As mentioned previously, in addition to reduced permeability, another advantage of utilizing blended cements in terms of improved durability is the reduction of CH contents in the hardened paste [1, 10]. From a calibration curve, the quantities of CH were determined from the averaged intensities of the 4.90, 2.63, and 1.80 Å diffraction peaks (Table 7).

Table 7 also shows the calculated CH content of the blended cement pastes, assuming that reductions were only due to dilution of the SRPC component (using the CH contents of the SRPC pastes at each age as references). However, from comparison of these two sets of data, it can be observed that FA (beyond 7 days curing for W/S = 0.25 and beyond 28 days for W/S = 0.36) and SF (at all ages) replacements resulted in substantial reductions in CH levels beyond those expected by dilution effects. In contrast, the BFS blends did not result in CH contents significantly different than those expected by dilution after all four curing periods. This observation, similar to that of Mehta [5] but different than that of Feldman [22], supports the argument that slag is hydraulic rather than pozzolanic in nature (that is, while slag requires an alkaline or thermally activated [23, 24] environment to hydrate, its own substantial calcium content is utilized to form CSH rather than consuming externally provided calcium).

TABLE 6—*Total intruded pore volume of pastes.*

Cementing Materials	W/S	Porosity Larger Than 0.0018 μm Radius, %			
		7 Days	28 Days	91 Days	182 Days
100% SRPC	0.25	14.4	14.9	11.1	11.5
35% FA	0.25	20.2	18.5	19.1	18.3
65% BFS	0.25	17.7	16.2	13.9	9.9
10% SF	0.25	15.4	12.6	11.7	10.7
20% SF	0.25	11.7	9.8	8.1	7.6
100% SRPC	0.36	27.2	25.1	23.7	21.9
25% FA	0.36	31.5	28.8	25.0	25.0
35% FA	0.36	32.0	29.7	29.2	25.2
50% BFS	0.36	27.5	25.0	20.4	16.9
65% BFS	0.36	29.3	25.0	21.4	20.2

Durability to Hot Brine

After 91 days moist curing at 23°C, paste prisms were immersed in either de-ionized water or a chloride/sulfate bearing groundwater at 70°C. Length and weight changes of the prisms are measured monthly. For pastes with W/S = 0.36, relative expansion measurements (expansion in the chloride/sulfate groundwater minus expansion in de-ionized water) up to 200 days exposure indicate that the SRPC paste is the most expansive (0.090% at 236 days) followed by the paste containing 25% FA (0.068% at 229 days), then the one with 50% BFS (0.039% at 200 days). The pastes containing higher replacement levels of FA and BFS have not expanded significantly (<0.027% at 168 days).

For W/S = 0.25, the only set of prisms exhibiting significant relative expansions was the SRPC paste. However, two of the three 20% SF paste prisms in the sulfate/chloride groundwater were found broken near the gage studs after only 23 days exposure. Since in early trials, the SF pastes were found to be especially sensitive to shrinkage cracking, it is thought that these prisms may have developed fine cracks while still restrained in the molds. This is supported by visible transverse cracking subsequently noted near the studs of the companion prisms in de-ionized water. The cracks would have allowed ingress of the chlorides to the bases of the stainless steel gage studs and corroded them, resulting in failure. Therefore, while there was no evidence of a lack of material durability, these broken prisms indicate the need to provide additional moisture very soon after casting specimens made with SF.

Discussion

Relationship Between Permeability and Durability

For the paste compositions made with W/S = 0.25, the higher expansion of the neat SRPC in the warm chloride/sulfate solution can be attributed to its higher

TABLE 7—Comparison of measured calcium hydroxide contents with values calculated from the dilution of portland cement.

Cementing Materials	W/S	Weight % SRPC in Blends	Calcium Hydroxide, Weight %							
			7 Days		28 Days		91 Days		182 Days	
			XRD	Calc[a]	XRD	Calc[a]	XRD	Calc[a]	XRD	Calc[a]
100% SRPC	0.25	100.0	16	16	18	18	17	17	19	19
35% FA[b]	0.25	71.8	10	11	8	13	7	12	6	14
65% BFS	0.25	36.7	8	6	6	7	7	6	7	7
10% SF	0.25	92.8	7	15	7	16	8	16	8	18
20% SF	0.25	85.1	4	13	2	15	0	14	0	16
100% SRPC	0.36	100.0	31	21	31	31	32	32	36	36
25% FA	0.36	80.4	26	26	24	25	24	26	21	29
35% FA	0.36	71.6	23	23	20	22	18	23	17	26
50% BFS	0.36	51.9	21	17	18	16	17	17	16	19
65% BFS	0.36	36.7	16	12	13	11	14	12	13	13

[a]Calculated by dilution of SRPC.
[b]% by volume replacement of SRPC.

TABLE 8 — *Critical pore radii of pastes.*

Cementing Materials	W/S	Critical Pore Radius,[a] μm			
		7 Days	28 Days	91 Days	182 Days
100% SRPC	0.25	0.0265	0.0235	0.0195	0.0165
35% FA	0.25	0.0345	0.0300	0.0255	0.0250
65% BFS	0.25	0.0280	0.0250	0.0210	0.0120
10% SF	0.25	0.0220	0.0190	0.0175	0.0180
20% SF	0.25	0.0140	0.0135	0.0120	0.0125
100% SRPC	0.36	0.0560	0.0490	0.0385	0.0350
25% FA	0.36	0.0740	0.0525	0.0420	0.0410
35% FA	0.36	0.0750	0.0515	0.0360	0.0350
50% BFS	0.36	0.0830	0.0445	0.0340	0.0275
65% BFS	0.36	0.0850	0.0470	0.0355	0.0350

[a]Corresponds to maximum on dV/dP versus log r plots.

permeability at 91 days (age of first exposure to the solution) and possibly its higher CH content. Likewise, the higher expansions of the neat SRPC made at W/S = 0.36 might be explained by its higher permeability after 91 days and its higher CH contents. However, the expansion of the prisms containing 25% FA cannot be explained by permeability. Therefore, based on Feldman's conclusions [22] and supported by the data shown in Table 7, it is postulated that insufficient FA was available to substantially reduce CH contents and provide resistance to this very aggressive exposure.

Relationship Between Permeability and Pore Structure

In some cases, it was found by others that permeability or durability was related to the volume of pores larger than specific radii such as 0.079 μm [25], 0.075 μm [26], 0.059 μm [9], and 0.030 μm [5]. (Pore radii were modified for a mercury contact angle of 140°). However, using the present data, none of these pore radii could be uniquely related to permeability.

Also, in other studies, the critical pore radius [14] — that is, the maximum of the differential pore size distribution curve (dV/dP) corresponding the maximum slope of the pore size distribution curve — was found to be related to permeability. The critical pore radius is similar to the maximum continuous pore radius described by Nyame and Illston [13] and the threshold diameter described by Winslow and Diamond [27]. However, in spite of the strongly linear relationship found between the log of permeability and the maximum continuous pore radius by Nyame and Illston [13], the reported relationship exhibited considerable scatter. For example, at a maximum continuous pore radius of 0.04 μm, their values of permeability ranged nearly three orders of magnitude from approximately 10^{-12} to 10^{-15} m/s. Values of the critical pore radius obtained in the present study are given in Table 8. No unique relationship was found between the log of permeability and critical pore radius; however, in most cases, linear relationships were found for each paste composition at the four ages of test. The sensitivity of permeability to the critical pore radius became less with paste compositions exhibiting smaller average critical pore radii. For example, the paste containing 20% SF had permeabilities reducing from 50 to $<1 \times 10^{-14}$ m/s, while critical pore radii only changed from 0.014 to 0.012 μm.

Therefore, it is concluded that either the pore size distributions of cement pastes are not adequately described by mercury intrusion (as suggested by Feldman for blended cements [12, 22] or other factors than pore size parameters affect permeability. Perhaps the density and CaO/SiO_2 ratio of the CSH affect permeability as well as strength [28]. In this regard, Mehta and Manmohan [25] found a better relationship between the log of permeability and the total porosity divided by the degree of hydration than from the volume of pores larger than a specific radius (0.079 μm corrected to a wetting angle of 140°).

Another explanation for the lack of relationship might be that mercury intrusion is performed on dried specimens while the permeabilities were measured on water saturated ones. This is important since it has been demonstrated that moisture content has a major effect on the microstructure of CSH, as water wedges into interlayer spaces [29]. Water displacement techniques result in higher densities and porosities than obtained with methanol or helium. Therefore, there is a definite possibility that the pore size distributions of saturated pastes could be much different than those obtained here.

In summary, no reliable method of accurately predicting permeability from other, more easily attainable properties is presently offered. However, the present study is still in progress and further analysis of the data may provide more specific conclusions.

Conclusions

The results indicate that sufficient replacements with supplementary cementing materials such as fly ash, blast-furnace slag, and silica fume reduce the ultimate permeability of a sulfate resistant portland cement paste. Of these materials, silica fume was most effective in reducing permeabilities at early ages. As well, the expansions of sulfate resistant portland cement pastes in strong chloride and sulfate solutions were reduced by the use of supplementary cementing materials, but these could not be explained by reduced permeability in all cases. While all three supplementary cementing materials reduced the calcium hydroxide contents of the pastes, 20% by volume silica fume was the most effective, eliminating calcium hydroxide completely after 91 days moist curing. From preliminary analysis, there does not appear to be an accurate way of predicting permeability from porosity or pore size parameters obtained by mercury intrusion for all of the pastes treated.

References

[1] Feldman, R. F. in *Proceedings,* Fifth International Symposium on Concrete Technology, Monterrey, 1981, pp. 263–288.

[2] Page, C. L., Short, N. R., and El Tarras, A., *Cement and Concrete Research,* Vol. 11, 1981, pp. 295–406.

[3] Emery, J. J. in *Extending Aggregate Resources, ASTM STP 774,* American Society for Testing and Materials, Philadelphia, 1982, pp. 95–118.

[4] Guyot, R., Ranc, R., and Varizat, A., "Comparison of the Resistance to Sulfate Solutions and Seawater on Different Portland Cements With or Without Secondary Constituents," American Concrete Institute, SP-79, Vol. 1, pp. 453–469.

[5] Mehta, P. K. in *Proceedings,* Fifth International Symposium on Concrete Technology, Monterrey, 1981, pp. 35–50.

[6] Sturrup, V. R., Hooton, R. D., and Clendenning, T. G., "Durability of Fly Ash Concrete," American Concrete Institute, SP-79, Vol. 1, 1983, pp. 71–86.

[7] Dunstan, E. R., *Cement, Concrete, and Aggregates,* Vol. 3, No. 2, 1981, pp. 101–104.

[8] Preese, C. M., "Corrosion of Steel in Concrete," SKBF/KBS Technical Report 82-19, Stockholm, 1982.

[9] Feldman, R. F. in *Proceedings,* Effects of Fly Ash Incorporation in Cement and Concrete, Materials Research Society, 1981, pp. 124–133.

[10] Manmohan, D. and Mehta, P. K., *Cement, Concrete, and Aggregates,* Vol. 3, No. 1, 1981, pp. 63–67.

[11] Mills, R. H., "The Permeability of Concrete for Reactor Containment Vessels," Atomic Energy Control Board of Canada Report INFO-0111, 1983.

[12] Gjorv, O. E. and Loland, K. E. in *Durability of Building Materials and Components, ASTM STP 691,* American Society for Testing and Materials, Philadelphia, 1980, pp. 410–422.

[13] Nyame, B. and Illston, J. M. in *Proceedings,* Seventh International Congress on the Chemistry of Cements, Paris, Vol. 3, 1980, pp. VI-181-185.

[14] Roy, D. M. and Parker, K. M., "Microstructures and Properties of Granulated Slag-Portland Cement Blends at Normal and Elevated Temperatures," American Concrete Institute, SP-79, Vol. 1., 1983, pp. 397–414.

[15] Feldman, R. F., *Journal of the American Ceramic Society,* Vol. 67, No. 1, 1984, pp. 30–33.

[16] Diamond, S., *Cement and Concrete Research,* Vol. 13, 1983, pp. 459–464.

[17] "Method of Test for Flow of Grout Mixtures (Flow Cone Method)," CRDC-C79-58, US Army Corps of Engineers (also Canadian Standards Association CAN3-A23.2-18, Clause 3).

[18] Kantro, D. L., *Cement, Concrete, and Aggregates,* Vol. 2, No. 2, 1980, pp. 95–105.

[19] MacInnis, C. and Nathawad, Y. R., *Durability of Building Materials and Components, ASTM STP 691,* American Society for Testing and Materials, Philadelphia, 1980, pp. 485–496.

[20] Marsh, B. K., "Relationships Between Engineering Properties and Microstructural Characteristics of Hardened Cement Paste Containing Pulverized Fuel Ash as a Partial Cement Replacement," Ph.D. thesis, The Hatfield Polytechnic, U.K., Feb. 1984.

[21] Nyame, B. and Illston, J. M., *Magazine of Concrete Research,* Vol. 33, No. 116, 1981, pp. 139–146.

[22] Feldman, R. F., "Significance of Porosity Measurements on Blended Cement Performance," American Concrete Institute, SP-79, Vol. 1, 1983, pp. 415–433.

[23] D'ans, J. and Eick, H., *Zement-Kalk-Gips* (in German), Vol. 7, 1954, pp. 449–459.

[24] Hooton, R. D. and Emery, J. J. in *Proceedings,* Seventh International Congress on the Chemistry of Cement, Paris, Vol. 2, 1980, pp. III-43–47.

[25] Manmohan, D. and Mehta, P. K. in *Proceedings,* Seventh International Congress on the Chemistry of Cement, Paris, Vol. 3, 1980, pp. VII-1–5.

[26] Goto, S. and Roy, D. M., *Cement and Concrete Research,* Vol. 11, 1981, pp. 575–579.

[27] Winslow, D. N. and Diamond, S., *Journal of Materials,* Vol. 5, 1970, pp. 564–585.

[28] Taylor, H. F. W., *Cement and Concrete Research,* Vol. 7, 1977, pp. 465–468.

[29] Feldman, R. F., *Cement Technology,* Vol. 3, No. 1, 1972, pp. 5–14.

Pierre Dutron[1]

Present Situation of Cement Standardization in Europe

REFERENCE: Dutron, P., **"Present Situation of Cement Standardization in Europe,"** *Blended Cements, ASTM STP 897,* G. Frohnsdorff, Ed., American Society for Testing and Materials, Philadelphia, 1986, pp. 144–153.

ABSTRACT: Since 1973 and at the request of the European Community which aims at removing trade barriers, the CEN (European Committee for Standardization) has undertaken to draft European standards (EN) for cements. Information is given on the progress of this work:

1. EN standards concerning test procedures and based on former International Organization for Standards (ISO) recommendations are for the major part ready for the final vote.

2. A draft nomenclature of cements, with particular emphasis being given to composite cements as a result of the energy crisis, has been drawn up while preliminary proposals for quality requirements are under consideration.

An analysis is also made of the contents of cement standards in European countries. Information is then given about the relative size of the current production of these four groups of cement—portland, portland composite (containing blast-furnace slag, natural pozzolana, fly ash, or even filler), blast-furnace, and pozzolanic—in Europe and in several other regions. Finally, possible developments in the types of cements to be produced between now and the end of the century are reviewed briefly.

KEY WORDS: portland cements, portland pozzolana cements, portland slag cements, slag cements, standards, performance, product specifications, production, Europe

Perhaps at no time in its history of over one hundred years has the cement industry found itself faced so suddenly with so many problems of vital importance, stemming largely from the oil crisis of 1973.

A major objective of the European industry naturally has been to save, by any possible means, some of the energy used in cement manufacture. This aim has been or is being steadily reached through two different courses, the effects of which are cumulative: (*a*) by reducing energy costs in the burning of clinker, and (*b*) by increasing the production of composite cements.

---

[1]Director, Cembureau, The European Cement Association, Paris, France.

Only the second course will be dealt with within the framework of this symposium. This involves replacing a proportion of clinker — high calorie consuming — by other products with similar properties which are suitable as such and do not require any further heat treatment. Such products already have been submitted to high temperatures naturally (for example, pozzolanas) or during manufacturing (for example, blast-furnace slag), or in the production of thermoelectric energy (for example, fly ash).

The production of composite cements is not a new development, and the concrete industry has had a long experience in the use of such cements. Industrialized production of cements with slag and with fly ash on the European continent started more than 50 and 25 years ago, respectively. The natural-pozzolana-based binders are much older. An accelerated development of composite cements has been witnessed over the past decade, mainly, for three reasons:

1. To save energy, as already stated.
2. Availability of large quantities of such constituents in many countries.
3. Environmental protection through the reutilization of industrial by-products combined with the preservation of the natural landscape through reduced quarrying of the raw materials needed for the production of clinker.

Present State of Standardization of Cement Categories in Europe

To allow such a development of composite cements, it is important that standards, codes of practice, and other regulations provide for all the categories of cements, including those containing secondary constituents, that can be produced by cement makers for the production of good quality concrete.

National Standards

Whenever necessary, national standards were revised and supplemented so as to include those categories of composite cements that can be produced in each country taking into account the availability of suitable natural materials or industrial by-products. This standardization work was carried out independently in each country, and, from an international point of view, led to some confusion as far as denominations are concerned. A similar product may be given different names; likewise, one denomination does not necessarily apply to the same product, depending on the country.

International Standards

This situation led the ISO Committee TC 74 on Cement to examine the matter as early as the 1960s. The work carried out resulted in the formal adoption of ISO Recommendation 597 on cement definitions and terminology. This was a commendable effort, although this document was ignored too often on the international level.

Later, the European Community issued a directive in 1969 aimed at removing barriers to the trade of products between member states. As cement was included in the list attached to the directive, the cement industry gave the matter considerable attention. It forwarded the results of its examination to the Commission of the European Communities which concluded that there were no barriers as such, although recognizing that trade would be easier, subject to three conditions being fulfilled. These concerned the standardization or harmonization of: (a) testing methods, (b) cement definition and nomenclature, and (c) a procedure for controlling quality and compliance with standards.

Against this background and at the request of a member state of the community, a Technical Committee TC 51 on Cement was set up by the European Committee for Standardization (CEN). It started work in 1973 and was soon given a working program by the Commission of the European Communities (EEC) which was quite close to the conclusions of the survey carried out by the cement industry.

Let us now examine how the matter stands and what are the prospects for a possible solution about ten years later.

Testing Methods — All the parties concerned, including the cement industry, agree that common standard test methods are of absolute necessity, that is, that harmonization should be total. This is justified from a scientific and commercial point of view and is essential for easier international recognition of quality control. This work is nearing completion, and the most important standards have reached the final voting stage.

All these standards will be referred to as EN 196 with the subdivisions shown in Table 1. Two parts of this standard call for some comments:

Part 1 — This is the confirmation on the European level of the so-called "plastic mortar" testing method, first known as the "RILEM-Cembureau method," which then became the International Organization for Standards (ISO) Recommendation 679. This was progressively accepted throughout Europe. It is recognized widely in Latin American countries, and its adoption as a standard is under consideration in other regions, including India and Australia.

In view of the importance of the accurate measurement of the mechanical strength of cement, Cembureau (European Cement Association) has published recently a manual in five languages which provides a complement to the European standard by drawing the attention of laboratories to details of the working procedure with relevant advice on checking the proper operation of the equipment.

To this must be added a view which, in our opinion, is of particular importance and represents the fundamentals of this standardization work. As a general rule, that is, except in specific cases, a testing method applies without change to any type of cement so that all cements are treated on an equal basis to make sure that the results are as comparable as possible.

TABLE 1 — *EN 196: testing methods for cement.*

Part	Heading	Chosen Method	Progress
1	mechanical strength	ISO R 679	final vote
2	chemical analysis	gravimetric analysis photometry complexometry	final vote
3	setting and soundness	vicat Le Châtelier	final vote
4	constituents content	dense liquid separation selective dissolution microscopy	preliminary vote
5	pozzolanicity test	ISO R 863	final vote
6	fineness	air permeability and sieving	draft
7	sampling	· · ·	draft

The application of this principle for specifying the water content of the mortar used for measuring the compressive strength required a considerable amount of work and discussions.

While there was unanimous agreement for applying the same rule to all cement types, choosing between a constant water/cement ratio (W/C) (as with portland cements in the United States), which treats all cements on an equal basis *vis-à-vis* their intrinsic value, and a constant workability (as with blended cements in the United States), which is more similar to site conditions, proved to be difficult.

It was feared that laying down a measurement of the workability, whose unavoidable scatter is well known, might produce an additional factor of inaccuracy, and this tipped the scales in favor of the first alternative.

Part 4 — As the growing use of composite cements makes it necessary that reliable measurement methods should be available to control the composition of such cements, that is, their clinker, slag, pozzolana, and fly ash contents, considerable work aimed at developing suitable methods was carried out, leading to a draft standard.

Cement Definition and Nomenclature — As already stated in the section on National Standards, standardization of this nomenclature on the international level was essential. As a result of long and exacting work, the draft proposal given in Table 2 was drawn up. Cements were classified into four main categories covering over 90% of the European production.

These categories are as follows:

1. *"Traditional" Portland Cement* — which may contain not only clinker but also up to 5% of other constituents such as slag, natural pozzolana, fly ash and fillers, the aim of such additions being essentially to improve rheological properties of the cement paste.

TABLE 2—*Draft CEN Standard 197-1.*

Category	Designation	Clinker	Granulated Blast-Furnace Slag	Proportion by Mass, % Pozzolana	Fly Ash	Filler
I	portland cement	95 to 100				
I				0 to 5		
II-S	portland slag cement	65 to 90	10 to 35		0 to 5	
II-Z	portland pozzolana cement	65 to 90	0 to 5[a]	10 to 35		0 to 5[a]
II-S/Z	portland composite cement	65 to 88	6 to 29	6 to 29		0 to 5
III	blast-furnace cement	20 to 64	36 to 80[b]		0 to 5	
IV	pozzolanic cement[c]	≥60	0 to 5[a]	≤40		0 to 5[a]

[a] The total content of blast-furnace slag and filler shall not exceed 5% by mass.

[b] Whenever the granulated blast-furnace slag content is higher than 65%, this shall be clearly indicated.

[c] Pozzolanic cement shall meet the requirement of the pozzolanicity test according to EN 196-5.

TABLE 3 — *Physical properties.*

Characteristic	Strength Grade[a]	Specification
Initial set, min min	32.5	60
	42.5	45
Final set, max h	all	12
Soundness, min mm	all	10

[a] See Table 5.

2. *Portland Cement with Secondary Constituents* — consisting of at least ⅔ clinker and a maximum of ⅓ of one or more of the aforementioned constituents. This category includes several subdivisions depending on whether the secondary constituent is slag (II-S), natural pozzolana or fly ash (II-Z), or two of these constituents at the same time (II-S/Z). In addition, it is possible in these three subcategories to add filler in a proportion limited to 5%.

In this respect, it will be noted that in the introduction to the CEN draft standard, a fourth category is mentioned under the name II-F. The only difference from II-S/Z is that the filler content allowed can be higher than 5%. As will be seen further on, such a cement is a standard cement in Spain, Finland, and France. In France, this cement is produced and used on a large scale, and the performance of structures using such cements is as satisfactory as when using more "traditional" cements. The CEN/TC 51 will examine in the near future whether technical conditions are fulfilled to include this cement in the future CEN standard.

3. *Blast-Furnace Cement* — already widely used in several countries, enabling the latent hydraulic properties of granulated blast-furnace slag to be utilized to a greater extent.

4. *Pozzolanic Cement* — which is of particular interest to Mediterranean countries where volcanic ash is widely available. It mainly differs from the II-Z cement in that it is required to pass a pozzolanicity test which enables the reactivity of the pozzolana used to be assessed.

Specifications

The program of TC 51 included a third facet, namely, the preparation of European standards on quality requirements. Undoubtedly, some degree of harmonization in this respect, for example, as far as testing ages for mechanical strength are concerned, is highly desirable. However, some people wonder whether full harmonization of all specifications is needed, particularly within the framework of the free movement of goods. Raw materials are different and so are weather conditions, which may justify strength levels not being the same everywhere. Nevertheless, at its last meeting (November 1983), TC 51 decided that the preparation of a full standard should be continued, and the proposals drawn up so far, which have not yet been given general approval, are shown in Tables 3 and 4.

TABLE 4—*Chemical properties.*

Characteristic	Cement Category	Strength Grade	Specification
Loss on ignition,	I, II-S, and III	all	5
max %	II-Z, II-S/Z, and IV	all	7
Insoluble residue,	I, II-S, and III	all	5
max %	II-Z, II-S/Z, and IV	all	...
SO_3,	I, II, and IV	32.5	3.5
max %	III	32.5	4
	all	42.5	4
Cl^-, max %	all	all	0.1[a]
Pozzolanicity	IV	all	EN 196-5

[a] A higher content is allowed whenever clearly indicated.

Tables 3 and 4 do not call for any comment other than to point up that only those characteristics of cement which need to be known in current applications have been included. When special properties are required, for example, alkali-aggregate reaction, heat of hydration, or even resistance to aggressive environment, reference shall be made to national standards.

In Table 5, only those values for the most widely used strength grades are specified, leaving it to national standards to define further grades either at a lower level, for example, for mass concrete, or at a high performance level, for example, for some prestressed concrete structures. The strength values shown in the table on the following page are to be taken with reserve as they could not be established with full knowledge of the facts until the details of the control procedure, dealt with hereafter, have been clearly defined.

While strength values may still be subject to some changes, the principle of specifying at 28 days not only a lower limit but also an upper limit is taken for granted. This is in line with a widespread approach on the European continent, the concept of a "bracket" already being applied in German, French, Greek, Dutch, and Swiss standards. This appeared to be the most efficient and simplest procedure for providing users with a guarantee of consistent strength in the cement supplied.

The data in Table 5 apply to any type of cement; any of the categories of cement in Table 2, theoretically, and often in practice, can fit into any strength grade.

Checking Compliance of Cement with Standards

The examination of this section has just been started within CEN/TC 51 and deals both with rules for the acceptance of a cement batch and the control procedure leading to the certification of cement.

The cement industry already has examined the latter for several years and met with two different types of difficulties:

TABLE 5 — *Proposed strength values,* N/mm^2.

Strength Grade	Subdivision	2 Days, min	7 Days, min	28 Days min	28 Days max
32.5	standard	· · ·	16	32.5	52.5
	early	8	· · ·	32.5	52.5
42.5	standard	8	· · ·	42.5	62.5
	early	16	· · ·	42.5	62.5

1. A central system of control by an outside body independent of the industry internal control and entitled to issue a quality label acknowledged by the authorities is not available in all countries. It is, therefore, difficult to convince those countries to set up such a system even when limiting its use to international exchanges.

2. The details of control procedures are, of course, different from one country to the next, and it is necessary to see that the levels of severity should be reasonably comparable.

Resumption of Work Within ISO

Standardization work at the European level has now reached a stage to allow ISO/TC 74 to resume its activities. To this end, it was agreed that whenever a CEN draft standard had gone beyond the stage of the preliminary vote, it would be submitted to the members of TC 74 for opinion. This was already the case with the methods for the measurement of setting, soundness, and mechanical strength. The results of the enquiry have shown that there is a large consensus for consideration of the work at the European level.

Distribution of Cement Production Between the Various Composition Categories

Let us now examine the present situation not only in Europe but also in some of the major cement producing regions on the basis of production statistical data, mainly, for 1982.

Table 6 shows the distribution, expressed in percentages, of cement production based on the four categories in the European draft standard. It is evident that the USSR and Eastern Europe, as well as India, have taken the lead, because in these countries portland cement now represents less than one third, and in some cases one tenth, of total cement production. Western Europe and China are placed favorably in second position, with cements in categories II to IV covering almost half of their production. Some Latin American countries follow close behind. By contrast, other continents still have a long way to go, because their production remains concentrated on portland cement, in most cases for at least 90% of production.

TABLE 6—Distribution of cement between the 4 CEN categories, as percentages of production.

Cement Categories		I	II					III	IV	Others
			II-S	II-Z	II-S/Z	Total			
Western Europe	Cembureau	53	6	6	13	25	7	13	2
Eastern Europe	Poland	10	57.5	0	6	63.5	26.5	0	0
	Yugoslavia	7.5	37	48.5	6.5	92	0.5	0	0
USSR		30	40.5	4	0	44.5	25.5	0	0
North America	Canada	100	0	0	0	0	0	0	0
	USA	97	2.5	0	0	2.5	0.5	0	0
Central America	Mexico	84	0	15	0	15	0	0	0
South America	Argentina	95.5	0	4.5	0	4.5	0	0	0
	Brazil	84	0	6	0	6	10	0	0
Africa	South Africa	93	4.5	0	0	4.5	2.5	0	0
Asia	India	35.5	0	47	0	47	17.5	0	0
	Japan	92	0	2	0	2	6	0	0
	China	55	38.4	1.7	0.7	40.8	2.4	0	1.8

TABLE 7—*Proportion of constituents other than clinker in cements, %.*

Years	1980	1990	2000
Western Europe	15	23	30
North America	3	12	20
Japan	1	10	20
OECD	9	16	25

It must be remembered, however, that within a region such as that covered by Cembureau Member countries, the situation may vary considerably from one country to another.

Thus, portland cement still reigns supreme in Ireland, Portugal, Sweden, Switzerland, and the United Kingdom, whereas at the other extreme, the share of portland cement accounts for a much lower percentage in a number of other countries: Netherlands (37%), France (27%), Austria (20%), Spain (19%), Greece (9%), and Luxembourg (6%).

This trend towards an increasing use of blended cements is gaining ground, particularly as a result of the development of the production of portland fly ash cement (II-Z), whether this cement is appearing on the market for the first time (as is the case of Denmark, the Netherlands, Belgium, Italy, and more recently Norway and Sweden), or the proportion of fly ash has increased, due to a revision of standards, from 15 to 20% to 30 or 35% (this is, or will be, the case in Austria and Spain).

In view of the impact which is made on energy savings by incorporating these various secondary constituents in cements, Cembureau conducted an enquiry for the International Energy Agency of the Organization for Economic Cooperation and Development (OECD) into the prospects for developments in the composition of cements from now until the year 2000. The results, which are of course only a guide, are summarized in Table 7.

It will be seen that there is still a long way to go between now and the year 2000, and it must be remembered that, taking into account the maximum proportions of secondary constituents technically permissible (about ⅓ for pozzolanas and fly ash and ⅔ for slag), cements with secondary constituents will eventually represent ¾, even ⅘ of production, and that consequently pure portland cements will be regarded as special cements reserved for applications where exceptional performance is required, particularly as far as mechanical strength is concerned.

Summary

The Symposium on Blended Cements sponsored by Committee C-1 brought about an exchange of information relevant to improvement of standards for blended hydraulic cements. Awareness of the need for improved standards for blended cements has grown steadily since 1973 as a result of the national need for energy conservation and the growing recognition that blended cements can offer benefits other than reduced costs for the manufacture of cement and concrete.

The nine papers in this volume bring together much information relevant to the development of standards for blended cements and mineral admixtures. It is noteworthy that several of the papers are from European authors whose different perspectives and knowledge should prove valuable to the ASTM committees. The volume begins with papers dealing with slag-containing cements and concretes, then presents a paper concerned exclusively with fly ash-containing cements, followed by papers dealing with more than one type of blending material or mineral admixture. It ends with a paper on the current status of the development of European standards for blended cements. Taking them in order, the papers may be summarized as follows.

The paper by Daube and Bakker reviews standards and applications for portland blast-furnace slag cements from the perspective of the BENELUX countries where slag-containing cements account for about half of all the cement used. Factors such as the composition and structure of the slag and portland cement clinker, which affect the quality of slag containing cements, are discussed with reference to ASTM, BENELUX, and British standards. The terms used for cements with different slag contents in the various standards are compared, and the special performance characteristics of portland blast-furnace slag cements and slag cements are reviewed. It is pointed up that, in the BENELUX countries, there is complete interchangeability between portland and portland blast-furnace slag cements, and the contractor may use any cement which complies with cement standards when there are no special technical requirements to be met. Slag cements offer benefits in applications where resistance to sulfates or seawater is required, or where there is a need for a cement with a low heat of hydration.

The paper on slag-containing cements by Frigione provides a good entry to the literature on these cements, as well as describing the author's research on cements in which slag is the predominant component. Frigione believes that the use of slag-containing cements will continue to grow. He presents data from a study in which he sought correlations between slag characteristics and compressive strength of mortar, sulfate resistance, resistance to alkali-aggregate reaction, and heat of hydration. He used ASTM test methods, wherever appropriate ones were available. In the case of sulfate resistance, the test methods used were the proposed ASTM method (now ASTM Method for Length Change of Hydraulic-Cement Mortars Exposed to a Mixed Sodium and Magnesium Sulfate Solution (C 1012-84)), and the method using small cubes exposed to a low-pH sulfate solution which was proposed by Mehta. The results show that resistance to sulfates and to alkali-aggregate reaction will be high if the slag content of the cement is high, irrespective of variations in the glass content of the slag, and the gypsum content and fineness of the blended cement. In general, it appears that slags containing some small amount of crystalline material perform better than fully glassy slags, possibly because of an effect of the crystalline material on the reactivity of the glass fraction or on the nucleation of hydration products.

In a paper presenting findings from three different, but related, projects with a practical orientation, Dubovoy, Gebler, Klieger, and Whiting discuss the effects of ground, granulated blast-furnace slags on properties of cement pastes, mortars, and concretes. They note that, for both mortars and concretes, there is an optimum level of replacement of portland cement by slag for which strength is maximized; this level is usually about 50% replacement. They also found that, at normal temperatures, early age strength development is retarded when slags are used, the extent of the retardation depending on the slag. The durability of air-entrained slag-cement concretes exposed to freezing and thawing in water is essentially the same as that of portland cement concretes, though their resistance to scaling by deicing salts appears to be somewhat less. The tests of Dubovoy et al show that ground, granulated blast-furnace slags can yield satisfactory concretes, whether the slag is used as a mineral admixture or as part of a blended cement, but the differences in performance between slags are sufficiently large that each slag should be characterized individually.

The use of the chemical shrinkage of a paste of portland cement and blast-furnace slag and the difference between its uptakes of water and another liquid, such as kerosene, to provide an easy method for the evaluation of the hydraulicity of the mixture is proposed by Mills. If, as Mills believes, there is reason to expect a functional relationship between strength and chemical shrinkage, then monitoring of chemical shrinkage might provide a new approach to nondestructive evaluation of concrete quality in the field. To illustrate the concept, Mills presents results on mixtures, each made from one of two portland cements and one of four slags. The mixtures were studied either as slurries which were continuously ball-milled in pycnometer bottles for up to three years, or as pastes cast into pycnometer bottles and compacted by vibration under vacuum; companion cubes of pastes, mortars, and concretes were cast for use in measurements of compressive strength. The chemical shrinkages of both types of specimen in pycnometer bottles were determined periodically. The ball-milled specimens were used to obtain data, such as the density of the hydration products and the mass of nonevaporable water per gram of cement at ultimate hydration, for use in interpreting the data on the cast specimens. Plots of compressive strength against the calculated volume concentration of hydration product were used to calculate the coefficients in an expression, $\sigma = AX^n$, relating strength to X, the volume concentration of hydration product. In other experiments, the volumes of kerosene taken up by water-saturated specimens which had been dried at $110°C$ were determined and compared with the volumes of water lost during the drying. The volume fraction, m, of evaporable water which resided in space not accessible to kerosene could be then calculated. While Mills found only a poor correlation between m and strength, he found good correlations between the 14-day values of m and drying creep and shrinkage. He concludes that m is a potentially useful parameter for estimating binder quality since, for the same porosity, a higher value of m indicates a higher volume of strongly-bound, relatively-immobile water and a lower permeance. On the other hand, drying creep and shrinkage increase with m because of the greater amount of "live" material. He further concludes that chemical shrinkage is a useful parameter for estimation of the volume concentration of hydration products. However, because the relationship between product concentration and strength is different for each cement, a separate calibration is needed for each cement.

Tenoutasse and Marion discuss the results of their research bearing on the mechanism of hydration of blended cements containing fly ash. From studies of the elements extracted from fly ashes by water, and by hydrochloric and hydrofluoric acids, they conclude that all of the sulfate and most of the potassium reside on the surfaces of ash particles, while almost all of the sodium is in the glassy phase. The effects on the particles of the treatments with water and acids were observed with the scanning electron microscope (SEM). Then, from studies of the pozzolanic reactions of the ashes with lime and with portland cement, they note that the lime etched the ash particles to give a product which, when viewed with the SEM, appears similar to that from the hydrofluoric acid treatment. With portland cement, the effects on the ash particles are also similar, but differences in the extent of attack suggest differences in reactivity between particles. At late ages, the ash particles are attacked sufficiently to leave the mullite crystals which were previously dispersed in the glassy phase. Measurements of the porosities and pore size distributions of hardened pastes of portland cement and fly ash show that replacement of cement by fly ash always increases the total porosity of the hardened paste, the effect being most obvious at early ages. By three months, however, the porosities and pore size distributions of pastes containing at least up to 25% of ash approach those of a cement paste without ash.

The problem of how to evaluate the performance of blast-furnace slags and fly ashes when blended or mixed with portland cement is addressed by Mills. He points up that, whereas reduced cost of the concrete is often the greatest incentive for using blending materials, the practitioner may have other good reasons for their use. These include improved resistance to sulfate attack and to alkali-aggregate reactions, and reduction of the seasonal variability of the cement. Mills outlines a procedure for designing concrete mixtures containing blending ingredients which will produce concrete of the same strength as the parent portland cement. The starting point is the establishment of appropriate boundary conditions, which may include the necessary target strength taking into account the variability of strength, the maturity to be achieved, the standard deviation of the strengths of concretes made with alternative cements, and the extra moist curing required for slow-hardening cements. In examples in the introduction to his paper, Mills uses 28-day strength as one boundary condition. If, as is usually the case, workability is also to be taken into account, the mixture must meet two interdependent criteria. An example is used to show that the cost of the binder required increases with both workability and strength. Similarly, increases in standard deviation of strength and workability must lead to increased cost as the cement content is increased to reduce the risk of failure. Mills recommends the use of efficiency factors for comparing the cementing qualities of materials used as partial substitutes for portland cement. He defines a "mass-strength efficiency factor" which takes account of strength, workability and characteristic variability, and a "maturity efficiency factor" which characterizes blended cements in terms of their responses to different curing regimes; the latter is defined in terms of the number of degree-hours needed for a concrete to attain the specified strength. Experimental data used to calculate efficiency factors for blast-furnace slags and fly ashes in different concrete mixtures are presented. The efficiency factor varies with the mixture design and is not a single-valued characteristic of a given mineral admixture. The implication is that concrete mixtures should be designed to optimize the appropriate efficiency factors for each application.

The effects attainable by intentional control of the particle size distributions of the ingredients in blended cements is discussed by Helmuth, Whiting, Dubovoy, Tang, and Love. In an earlier study, they had shown that controlled particle size distribution (CPSD) portland cements have energy-saving potential because, by suitable choice of the size distribution, a given level of performance can be obtained at a lower specific surface area and with the expenditure of less energy in grinding. In the case of blended cements, they also show that there are significant benefits to be gained by control of particle size distribution, particularly in increasing early age strength development. Their initial studies were of pastes of 45 blended cements. These cements were made with one cement selected from two CPSD portland cements and a normally ground portland cement, and selections from six different powdered mineral admixtures (two fly ashes, one Class F and one Class C; one ground, granulated slag; one coarser slag; one silica fume; and a ground limestone), two water-reducing admixtures, and one accelerating admixture. Subsequently, based on the large amount of data for the pastes, they selected five of the blended cements for further testing as mortars and concretes. The concrete tests were of mechanical properties (compressive and flexural strength and elastic modulus), drying shrinkage, and sulfate resistance. The final conclusions are that cement pastes, mortars, and concretes made with CPSD blended cements have properties which are approximately equal or superior to those made with normally ground blended cements of the same compositions. The major benefit in the use of CPSD blended cements may be to produce concretes with early age properties comparable to those obtained with portland cements, thereby making them more readily acceptable to users.

Many aspects of the performance of a concrete are intimately linked to the pore structure and permeability of the cement paste matrix. Because the permeability is particularly important in concretes for use in containment of radioactive wastes, Hooton has sought relationships between pore structure and permeability of portland cement pastes containing fly ash, slag, and silica fume as a step towards reducing the need for permeability measurements. His results show that replacement of a sulfate-resistant portland cement with any of fly ash, slag, and silica fume reduces the ultimate permeability to water of a portland cement paste. Silica fume is particularly effective in reducing permeability at early ages. Because the resistance of the blended cement pastes to aggressive environments cannot be explained solely in terms of differences in permeability, it appears that reduction of the calcium hydroxide content of the hardened paste by pozzolanic action is also important. Silica fume is most effective in reducing the calcium hydroxide levels, and slag the least effective. For the cement used, 20% of silica fume is sufficient to eliminate calcium hydroxide completely in 91 days of moist curing. Hooton's preliminary analyses suggest that there is no accurate way of predicting permeability from porosity or pore size parameters determined by mercury intrusion.

The last paper is of special interest from the standards viewpoint. It is a review of the present situation regarding cement standardization in Europe. The author, P. Dutron, is chairman of the European Committee for Standardization of Cements, CEN TC 51. He analyses the contents of cement standards in several European countries and describes the six categories of cements — portland cement and five categories of blended cements (composite cements) — defined in the draft CEN Standard 197-1. Each category of blended cement may contain at least 5% of other blending materials than those indicated in its name. For example, portland pozzolana cement, Category II-Z, must contain 65 to 90% of clinker and 10 to 35% of fly ash or pozzolana; it may also contain up to a combined total of 5% of blast-furnace slag and filler. In general, the principal of specifying 28-day strengths with both upper and lower limits "is taken for granted." Now that the CEN standard has gone beyond the preliminary voting stage, it is to be submitted to ISO Committee TC 74. The paper gives figures showing the large differences in some national productions of cements in the categories proposed in the draft CEN cement standard. The author predicts that, between now and the year 2000, the use of blending ingredients will increase substantially in many regions of the world, including North America. Eventually,

Author Index

Subject Index